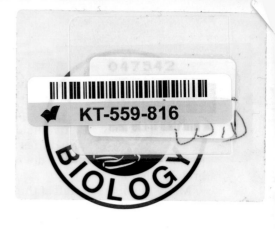

PLANT SCIENCE IN ACTION

Caroline Barnes
and Nick Poore

for the
University of Nottingham

Hodder & Stoughton

A MEMBER OF THE HODDER HEADLINE GROUP

Acknowledgements

The original idea to increase awareness of agriculture and horticulture as applied biological sciences in schools and colleges was conceived through discussions between the National Farmers Union and the University of Nottingham, Department of Agriculture and Horticulture, both of whom initially financed the venture. Invaluable support was also given by Bancroft's School, Essex in seconding Nick Poore from his teaching duties to start writing this text, and by the LSA Charitable Trust who funded Caroline Barnes to further develop and complete this book. All of these organisations have our grateful thanks.

Many members of staff at the University freely gave data, observations and numerous examples for inclusion in this text. Particular thanks are due to Keith Scott, Dave Stokes, Charles Wright, Peter Alderson, Sayed Azam-Ali, Jerry Roberts and Ian Taylor. Thanks must also go to Dr Susan Brough for her valuable information about The Leaf for Life Project, Dr B. Mulligan and Dr M. Anderson of the *Arabidopsis* Stock Centre, Nottingham University, Mr R. Price of SAPS (Science and Plants in Schools) based at Homerton College, Cambridge and Mr P. Freeland of Worth School, Crawley for help in the writing of several investigations. Many thanks, also, to Christopher Barnes for his invaluable help and advice with the agricultural content of this book.

For Christopher, Matthew and Oliver.

Nick Poore would like to thank the Headmaster and Governors of Bancroft's School for their support.

British Library Cataloguing in Publication Data

ISBN 0 340 60099 3

First published 1994
Impression number   10 9 8 7 6 5 4 3 2 1
Year   1998 1997 1996 1995 1994

Copyright © 1994 The University of Nottingham.

Typeset by Litho Link Ltd, Welshpool, Powys, Wales.
Printed in Great Britain for Hodder & Stoughton Educational, a division of Hodder Headline plc, 338 Euston Road, London NW1 3BH, by Bath Press Group, Bath.

CONTENTS

PREFACE

The Agricultural and Horticultural industries rely heavily on a knowledge and understanding of plant physiology. This understanding allows growers to manipulate plants and so produce food efficiently. Agriculture is concerned with the cultivation and production of the crop species which, in general, provide the basic components of the human diet, for example wheat, rice, potatoes. Horticulture involves more intensive techniques and is concerned with the production of more expensive species, for example fruit, flowers and the more exotic vegetables. This book is about how we have used our knowledge of science to develop crop species of all kinds and to improve their productivity.

We start by examining the important relationship we have with plants as sources of food, of fuel and of many other essential products.

We talk about productivity and yield, as we consider the environmental factors which affect the production of plants for food and the ways in which these can be manipulated to increase productivity. We look at pests and diseases which have a great effect on crop yield and discuss methods of reducing crop losses.

We also discuss how modern agricultural practices affect the environment, and see how legislation produced by the EC and other bodies helps to maintain stable market prices for agricultural produce, and to reduce surpluses.

Special consideration is given to the use of protective environments (greenhouses) and the technology associated with the production of luxury produce by the horticultural industry.

Methods used by growers to improve crop plants through breeding programmes are reviewed, and then we go on to look at the 'Green' Revolution and its impact on developing nations.

Finally, we look towards the future with a review of some of the alternative food sources currently being developed in order to help feed the world's ever increasing population.

This book aims to cover the materials found in the applied biology or related option modules offered by many examining boards in Biology A level, but the material is also suitable for students following related courses at BTEC and similar levels. General Studies courses may also draw heavily on this text.

It is expected that students already have a knowledge of basic plant physiology and anatomy before studying the option modules for which this book is relevant. In particular a knowledge of the following would be an advantage:

- flower structure
- pollination and fertilisation in flowering plants
- seed and fruit formation, germination
- photosynthesis
- transpiration
- basic Mendelian genetics.

1 PLANTS FOR FOOD

LEARNING OUTCOMES

After studying this chapter you should be able to:

- appreciate the impact of plants on our daily lives,
- describe seed formation and explain the roles played by both temperature and light in the breaking of seed dormancy and germination,
- describe the importance of plant hormones in the formation of fruit and flowers,
- define the meaning of vegetative propagation and distinguish between plant structures designed for this purpose.

1.1   THE IMPORTANCE OF PLANTS

Plants have many, varied and important influences on our lives. The most important use we make of plants is as a source of food either directly as things we consume ourselves like fruit, vegetables, cereals or indirectly as foodstuff for the animals that we eat.

. . . or indirectly

Plants play an essential role in the balance of the natural environment. During photosynthesis they remove potentially harmful carbon dioxide from the atmosphere and replenish the air with oxygen. They provide habitats for a huge variety of wildlife, and provide us with a source of beauty and pleasure.

Plants provide habitats

The wood we use for building, and pulp we use for paper production all come from plants. They provide us with fuels in the form of coal, and wood. From oil, a fossil fuel made largely from plant material, man has developed many synthetic products like tar, plastic and petrol.

Plants provide food directly . . .

1

Plants provide fuels such as coal

Plants are a source of natural fibres like cotton and rubber, of many dyes, and of medicines and perfumes. This book explores our relationship with just one group of plants — the food crops.

1.1.1 Plants as autotrophs

Plants are **autotrophs**, which means that they are able to manufacture their own food using only carbon dioxide, water and solar energy. They do this by the process of **photosynthesis** which is discussed in Chapter 2. Photosynthesis enables plants to manufacture energy — rich organic compounds which they use to develop and grow, but can also be used by animals, including humans, for food. We rely on plants to provide *all* the food we need either directly, or indirectly by producing feed for the animals we eat as meat.

Over the course of history, more than 3000 plant species have been used as food by humans. However, we have selected only a handful of these species for **domestication**. These common 'domestic' food crops have been subjected to highly selective breeding programmes creating high yielding, good quality varieties.

Only a small number of plants are commonly used for food today

1.1.2 Plants provide us with food

Plants directly provide 88% of the energy and 80% of the protein in the human diet worldwide. The actual proportions of dietary energy and protein provided by plants varies from country to country as shown in Table 1.1.

Table 1.1   Dietary energy and protein provided by plants for the human population

REGION	PERCENTAGE OF ENERGY PROVIDED BY PLANTS	PERCENTAGE OF PROTEIN PROVIDED BY PLANTS
Asia	95	78
Africa	93	79
Europe	81	53
Latin America	85	76
Oceania (Australia, New Zealand, Polynesia)	65	30
North America	68	30
World	80	88

(From: *Plants, Food and People* by Chrispeels and Sadava. Copyright © 1977 by W. H. Freeman, reprinted with permission.)

Table 1.1 shows us that highly developed industrialised regions, such as Europe, Oceania and North America rely heavily on both plants and animals for their dietary energy and protein, whereas less developed regions rely almost entirely on plants for their food. This is because using animals for food is much more expensive and less energy efficient than using plants directly. When an animal is fed plant material, a certain proportion of the plant's energy and protein content is used by the animal during its metabolic processes, like respiration. So some of the energy contained in the plants an animal eats is used up and not stored as muscle (see Figure 1). Because energy is lost at each level of a food chain in this way, it is more energy efficient for a person to eat plant food directly, than to feed livestock and eat the resulting meat.

We use many different parts of the plant to provide us with food. Different plant structures provide different proportions of nutrients and their composition often reflects their function within the plant as a whole.

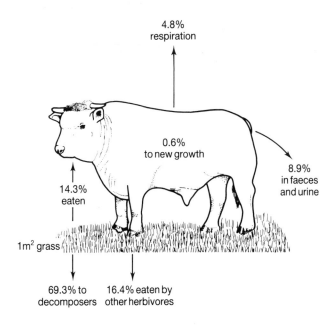

4.8%
respiration

0.6%
to new growth

8.9%
in faeces
and urine

14.3%
eaten

1m² grass

69.3% to
decomposers

16.4% eaten by
other herbivores

Figure 1   The fate of the energy from a year's growth of 1 m² of grass grazed by one bullock

1.2   SEEDS

Seeds develop from the ovule which is formed deep within the flower. The ovule contains the female gamete – the ova – and once fertilised will develop into the embryo of the seed. It is the seed embryo which will develop into a new plant.

The fruit develops from the ovary wall and surrounds the seed. A fruit often helps the dispersal of the seed as its taste and colour encourage animals and insects to eat it. The seed is then carried inside the animal and deposited later, away from the parent plant.

Both seeds and fruit are important to humans as food sources. We rely on many seeds for our staple foods: cereal grains such as rice, wheat and barley, and legumes like peas, lentils and beans, form the basis of most human diets.

1.2.1.   Seed formation

Let us recap what we know about the formation of seeds in flowering plants following fertilisation. During fertilisation the male gamete (the **pollen generative nucleus**) and the female gamete (the **ovum**) combine to produce an embryo.

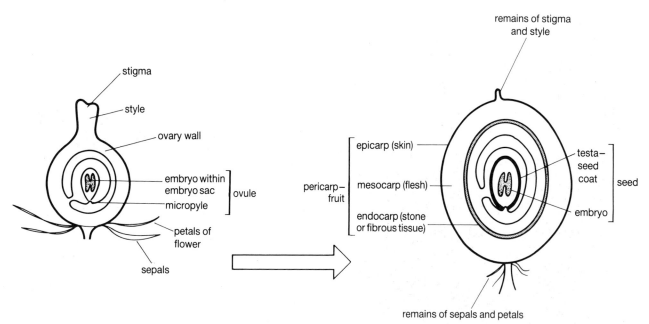

stigma

style

ovary wall

embryo within
embryo sac

micropyle

ovule

petals of
flower

sepals

remains of stigma
and style

epicarp (skin)

pericarp –
fruit

mesocarp (flesh)

endocarp (stone
or fibrous tissue)

testa –
seed
coat

seed

embryo

remains of sepals and petals

After fertilisation the *embryo* develops
within the embryo sac.
The ovule develops into the *seed* and
the ovary wall swells to become the *fruit*.

Figure 2

3

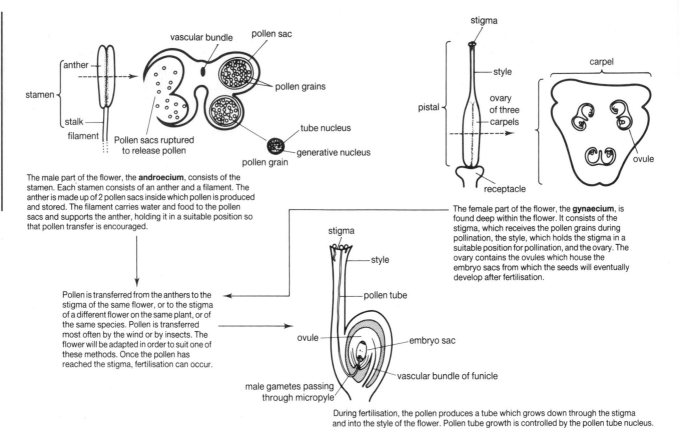

The male part of the flower, the **androecium**, consists of the stamen. Each stamen consists of an anther and a filament. The anther is made up of 2 pollen sacs inside which pollen is produced and stored. The filament carries water and food to the pollen sacs and supports the anther, holding it in a suitable position so that pollen transfer is encouraged.

Pollen is transferred from the anthers to the stigma of the same flower, or to the stigma of a different flower on the same plant, or of the same species. Pollen is transferred most often by the wind or by insects. The flower will be adapted in order to suit one of these methods. Once the pollen has reached the stigma, fertilisation can occur.

The female part of the flower, the **gynaecium**, is found deep within the flower. It consists of the stigma, which receives the pollen grains during pollination, the style, which holds the stigma in a suitable position for pollination, and the ovary. The ovary contains the ovules which house the embryo sacs from which the seeds will eventually develop after fertilisation.

During fertilisation, the pollen produces a tube which grows down through the stigma and into the style of the flower. Pollen tube growth is controlled by the pollen tube nucleus. As the tube grows down through the style, the generative nucleus will divide by mitosis and form two male gametes. When the tube tip reaches the bottom of the style, it enters the embryo sac by growing in through an opening (the micropyle). The pollen tube nucleus will disintegrate and one of the male gametes will fuse with the female ova.

Look at Figure 3. **Pollen** is produced in the male part of the flower, by the **anthers**. A mass of **pollen mother cells** develop inside each lobe of the anthers. These cells will eventually produce the **haploid pollen grains**. Each pollen grain nucleus divides by **mitosis** to produce a generative cell nucleus (the male gamete) and a **pollen tube nucleus**.

Figure 3   Gamete production and fertilisation in the flowering plant

The female gametes (the ova), are found deep within the flower inside the ovule. **Embryo sacs** develop inside the ovule, each of which contains eight haploid nuclei. Just one of the nuclei will develop into the egg cell (ovum). When the ovum is fertilised the ovule develops to form the seed, inside which the embryo is located.

1.2.2 Seed structure

The structure of a seed is adapted for survival through times when environmental conditions are unfavourable for growth. Once formed, seeds often dry out, become dormant and development stops — they may survive in this form for months or years. When conditions change and become favourable for growth, the embryo within the seed can then start to grow. As it has no leaves for photosynthesis, the seed relies upon the food stored within its endosperm for the nutrition it needs for early growth. It is this nutritious endosperm tissue which makes seeds such a good source of food for us and other animals.

Figure 4 shows the structure of a typical cereal seed and a typical legume seed.

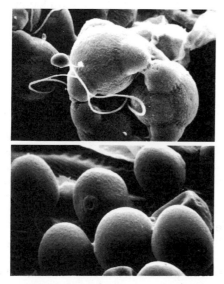

Pollen grains (top) Timothy Grass and (bottom) Rosebay Willowherb

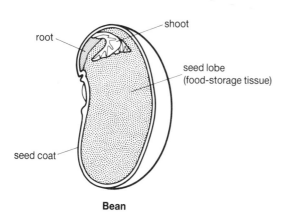

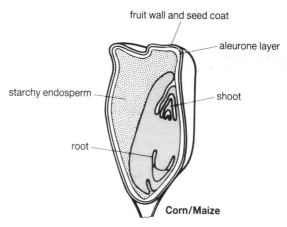

Figure 4 *The structure of a leguminous and a cereal seed*

As you can see cereal grains, like wheat, rice and barley, are surrounded by a protective seed coat (the **testa**), which is composed of indigestible fibre. The embryo, consisting of the **plumule** (which develops into the shoot), and the **radicle** (which develops into the root), is typically rich in protein, fat, sugars and vitamins B and E. The **endosperm** takes up most of the seed and contains most of the starch and protein. It is the endosperm which provides the germinating seed with nutrients. The endosperm is surrounded by the **aleurone layer**, the importance of which we shall discuss later. Cereal seeds contain between 8–14% protein and 70–75% carbohydrate.

Leguminous seeds, like peas and beans, contain considerably more protein than cereals (15–40%) and less carbohydrate (20–50%). The seed is, again, surrounded by the protective testa and the food for the germinating seed is stored in the seed lobe. Legume seeds provide an important source of protein in the diets of many people (especially vegetarians and those living in less developed areas of the world). Seeds from legumes contain a lot of protein because leguminous plants have bacteria living in nodules on their roots, which are able to 'fix' atmospheric nitrogen as nitrate. The nitrate can be absorbed by the plant and used to make protein, some of which is stored in seed lobes to provide for germination and growth. The nitrogen fixing ability of legumes means they can be grown in soils which are nitrogen-poor and even add nitrogen to the soil for use by successive non-leguminous crops.

1.2.3 Sowing seeds for new crops

In agriculture, a proportion of the harvested seed from any crop must be saved to be sown in subsequent years. Seeds which have been saved to produce new crops:

- must be viable, i.e. the embryo must be alive,
- must pass through any necessary dormancy-breaking processes and experience suitable environmental conditions to trigger and sustain growth.

These conditions which trigger and sustain growth are of particular interest to farmers and commercial growers, who are looking for good **field establishment**. This means they expect a high percentage of the sown seed to germinate, and want synchronous, predictable and rapid emergence of the seedling in the field. To achieve good field establishment, a farmer must use good quality seed, certified by the Ministry of Agriculture, Fisheries and Food (MAFF).

1.2.4 Seed viability

Seed viability is of great importance to the grower. The cost of seed will generally be a very small proportion of the cost of growing the crop and so it makes economic sense for growers to be prepared to invest in good quality living seed from certified seed merchants. Seed merchants need to maintain confidence in their product and so they apply rigorous criteria, with legally enforced minimum standards, to control quality. There are also strict rules to ensure consistency in the testing methods. These involve the methods of sampling the seed batch, as well as the conditions under which the tests are carried out.

1.2.5 Measuring seed viability

Seed viability can be measured in several ways:
- **Germination tests**. Seeds absorb water from moist filter paper in petri dishes. **Percentage germination** can then be calculated from the number of seeds which successfully germinate. Each crop species has specific conditions under which this test should take place. These relate to the amount of water used in the petri dish, the amount of contact between seeds and water, the grade and number of sheets of filter paper, the temperature, illumination and, very importantly, the duration of the test.

Germination tests give the seed merchant information about the percentage of seeds which will germinate within a fixed period of time — usually about 28 days. They also give information about the length of time it takes for most of the seeds to germinate. The speed of germination is a crude measure of the seeds' vigour. Vigorous seed will reach a given percentage germination (say 75%) much more quickly than less vigorous seed.

- **The tetrazolium test** is a chemical test. Seeds are soaked in water for 24 hours and then cut in half and transferred to a 1% solution of tetrazolium chloride. After four hours the cut surface is examined. Because active mitrochondria are dyed red by the tetrazolium chloride the cut surfaces of seeds which are viable, will appear red in colour. Seeds which are not viable will have no active mitochondria to take up the solution and so appear neutral.

- **The catalase test** is a qualitative test. Seeds are placed into specimen tubes with a solution of 1% sodium hypochlorite solution. This solution kills any micro-organisms on the seed surface. The seeds are then washed and soaked in a very dilute solution of hydrogen peroxide for six hours. The enzyme **catalase**, which is found within viable seeds but not in seeds which are not viable will break down the hydrogen peroxide releasing oxygen and water. As the water is released the concentration of the solution will fall. So by measuring the concentration of the hydrogen peroxide solution periodically using Merck Antherperoxide reagent strips, it is possible to assess whether seeds are viable or not. In general viable seeds will completely break down the peroxide present, to give water alone, in 6–12 hours.

Seed vigour results from a combination of genetic and physical characteristics, such as the age of the seed, the average seed size or weight and whether or not it is carrying seed-borne pathogens such as fungi causing smuts (*Ustilaginales*).

> **Now try Investigation 1 Measuring Seed Viability in the *Plant Science in Action Investigation Pack*.**

1.2.6 Seed dormancy

It is an advantage to the seed to remain dormant after fertilisation. In nature seeds usually form towards the end of a warm summer season, so it is better to wait out the unfavourable conditions of the coming winter in a dormant state, and be triggered to grow when conditions improve the following spring. Seed dormancy can be broken, and growth triggered by a variety of mechanisms, depending on the species involved.

(a) Stratification (Chilling)

Stratification is the term used to describe the chilling of seeds to stimulate germination. Many types of seed, like peach, apple and maple, germinate in response to chilling. Only after exposure to low temperatures for a period of a few weeks, will they germinate. Under natural conditions, the chilling occurs over winter and the seeds then germinate in springtime.

Chilling appears to affect the hormone concentration inside the seeds of some species. We think that the **abscissic acid** contained in the seed coat or **testa** inhibits the production of the **giberellins** which are required for germination. After a prolonged period of cold temperatures the abscissic acid level within the seeds of sensitive species, like apple, peach and maple, falls and so the production of the giberellins is no longer inhibited. The giberellins produced by the embryo will stimulate germination. This is why in many species, seeds will only germinate in the spring after they have experienced the cold winter temperature. The relative concentrations of these two chemicals over time, and their effects on germination in maple seeds can be seen from Figure 5.

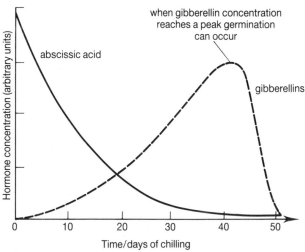

Figure 5   Graph to show the effects of chilling on the germination of maple seeds

(b) Light

In some species light is required to stimulate seed germination. Most seeds *can* germinate in total darkness, but some species produce light-sensitive seeds, like lettuce for example. The response to light is controlled by the **phytochrome system**. This system uses a pigment called **phytochrome** (which is quite separate from the photosynthetic pigment, chlorophyll). Phytochrome is a blue-green pigment, found in very small quantities (0.1 parts per million) in all plants, mainly in the leaves. It controls many biochemical processes including flowering and seed germination.

Phytochrome occurs in two interchangeable forms or **isomers**: the P_R (**cis** isomer) absorbs red light (which has a wavelength of 665 nm) and the P_{FR} (**trans** isomer) absorbs far-red or infra-red light (wavelength 725 nm). When P_R absorbs red light, it rapidly changes into P_{FR} and when P_{FR} absorbs far-red light, it rapidly changes into P_R. P_{FR} will also slowly change into P_R in the dark. In sensitive species, like lettuce, the relative proportions of these two forms of phytochrome determine whether or not dormancy is broken and germination can occur.

Table 1.2 shows the percentage germination in seed samples of one lettuce species in response to various light treatments

LIGHT TREATMENT IN 24 HOURS	PERCENTAGE GERMINATION
R	70
R/FR	6
R/FR/R	76
R/FR/R/FR	6
R/FR/R/FR/R	76
R/FR/R/FR/R/FR	7
FR/R/FR/R/FR/R	81
R/FR/R/FR/R/FR/R/FR	7

R = red light (665 nm)   FR = far-red light (725 nm)

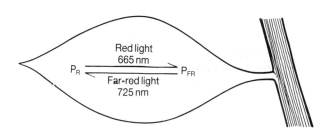

Figure 6   The interaction between phytochrome pigments in plants

This particular species of lettuce appears to be sensitive to changes in wavelength of light. Red and far-red light is detected by the phytochrome system as discussed above. It seems from the results that, regardless of the initial stages of light treatment it is the final exposure to red light that stimulates germination in this lettuce variety. Red light will bring about the conversion of the phytochrome form P_R to P_{FR} and so high concentrations of P_{FR} must be required for germination to occur in this species.

> **Now try Investigation 2 The Effects of Light on the Germination of Lettuce Seeds in the *Plant Science in Action Investigation Pack*.**

1.2.7 Germination and the mobilisation of food reserves

Germination can occur once dormancy has been broken. The stages of germination are summarised in Figure 7.

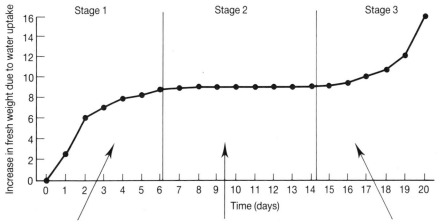

During the first stage, seeds take up water. Initially this is by imbibition through the micropyle and testa. The testa contains hydrophyllic colloids which absorb water rapidly, causing them to swell and eventually the testa ruptures. Water is then able to enter the rest of the seed, moving into the embryo and endosperm by diffusion and activating enzymes.

During the second stage the rate of water uptake slows down as the seed becomes fully imbibed but there is a resumption of metabolic activity, especially in the embryo. In most seeds, the embryo is fairly close to the surface and so it receives water and oxygen promptly and is able to become metabolically active very early in germination. Hydrolytic enzymes and growth regulators are produced by the embryo during this phase.

During the final stage the radicle begins to emerge. Water uptake is resumed, partly due to the effect of hydrolytic enzymes increasing the soluble sugar content of the seed and partly due to rapid water uptake vacuolating radicle cells. This phase continues until germination and emergence are complete, i.e. when the seedling is established with a root system anchoring the plant and absorbing water and minerals and with a shoot system supporting photosynthesising leaves or cotyledons.

Figure 7   The phases of seed germination

The most important stage of germination is when the seed takes up water. This allows the testa to soften, the endosperm to swell and so split the testa. Water also reactivates enzymes within the endosperm. These enzymes cause the breakdown and mobilisation of food stored in the seed, so providing nutrition for the developing embryo.

In most seeds, these processes are very largely under environmental control. In both leguminous and cereal seeds there are food storage cells which are living and produce **hydrolytic enzymes**. These hydrolytic enzymes digest the stored foods in the seed, releasing soluble compounds which can be taken by the embryo. As happens in our own intestine hydrolytic enzymes catalyse the breakdown of complex molecules into simpler molecules which can be used in cells. The rate at which this breakdown takes place in plants seems to be dependent on environmental factors like the ambient temperature and the oxygen supply.

However there is some evidence, though that the embryo itself might have a certain amount of control over the mobilisation of stored food. For example, in cereals, the bulk of the endosperm is non-living and unable to produce hydrolytic enzymes. But a thin layer of endosperm tissue, just beneath the testa, remains alive and stores large quantities of protein. This is the aleurone layer. In response to gibberellic acid secreted by the germinating embryo, the aleurone layer synthesises and releases amylase and maltase enzymes. These two enzymes pass into, and hydrolyse the starchy endosperm, releasing soluble sugars which are then absorbed by the embryo. This example of how the embryo itself has some control over the release of food stored in the seed is summarised in Figure 8.

Now try Investigation 3 Investigating Aleurone Activity in Barley in the *Plant Science in Action Investigation Pack.*

1.3 FRUITS AND FLOWERS

Many crop plants are grown for their fruits or flowers. Fruit is often a good source of minerals and vitamins and some fruits, like bananas, provide both carbohydrate and protein as well.

We eat some flowers as vegetables, such as, cauliflowers and broccoli, although unlike seeds and fruits the main function of the flower as far as the plant is concerned is not to provide nourishment.

1.3.1 The development of fruit

A seed that develops from the ovule is often surrounded by a fruit. The fruit is formed from the ovary of the flower. Fruit formation is controlled by the plant hormone **auxin**. For fruit to develop or be 'set', auxin must be produced by the ovule and ovary after fertilisation. If this does not occur, the flower will drop. This is called **abscission**, and fruit will not form.

Auxins are used commercially in fruit production in two ways. Firstly, to enhance growth and make bigger final fruits, in tomato production for example, as shown in the photograph below. Secondly, auxins can be used to produce 'seedless' fruits which are sometimes popular especially in developing countries. Some species develop fruits naturally without fertilisation so the fruits contain no seed. This is called **parthenocarpy**, and the process can be induced artificially by adding auxins, in species like tomato, squash and pepper. The same effects can be induced in other fruits, like cherry and grape, when another hormone called gibberellic acid is applied in the appropriate way.

GA = giberellic acid

Figure 8   The mobilisation of food reserves in barley

Auxins can be used to enhance fruit growth in many species.

These techniques have resulted in the availability of both seeded and seedless varieties of some common fruit and vegetables, increasing consumer choice. Many consumers prefer seedless varieties and are willing to pay a higher price for these products.

Some examples of seedless fruit

Fruit ripening is often accompanied by a burst of rapid respiration. This burst is called **climacteric** respiration and, in the case of bananas and avocados is induced by a sudden rise in the level of ethylene within the fruit. But not all fruits are climacteric: oranges for example show no change in the rate of respiration as they ripen. Figure 9 shows the different patterns of respiration in climacteric and non-climacteric fruits during ripening.

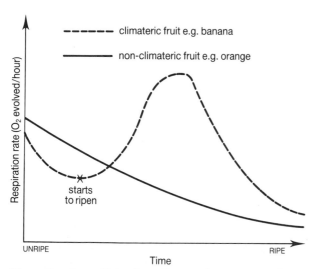

Figure 9 A graph to show the respiratory rates in climacteric, and non-climacteric fruits

1.4 VEGETATIVE STRUCTURES

The **vegetative structures** of a plant are structures which are not involved in sexual reproduction and seed production. Stems, leaves and roots for example are all vegetative structures.

These structures are clearly not designed specifically to provide food, unlike seeds and nuts, nevertheless there are many crop species which are grown in order to harvest stems, leaves and roots for food.

The most economically important stem crop is sugar-cane. Sugar-cane is a tropical crop which produces sucrose by photosynthesis and stores it in its stem. Sugar-cane provides 65% of the world's sugar, which is extracted by crushing the stalks, squeezing out the juice and boiling. The water evaporates as steam and the liquid becomes concentrated. This liquid is then refined for the Western market by crystallisation. The waste product after crystals of pure sugar have formed is called **molasses**, which is used in rum production or as animal feed.

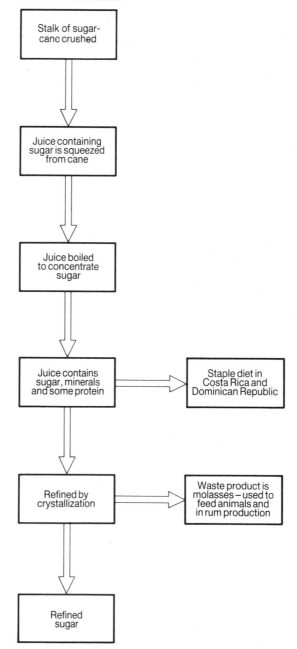

Figure 10 The production of refined sugar

Sugar-cane growing in India

A number of plant species are grown for the nutritive qualities of another vegetative structure — their leaves. Leafy vegetables, like lettuce and spinach for example, provide large amounts of vitamins and minerals for both human and animal consumption.

Root crops are of enormous importance as a food source. A wide variety of root crops are grown and eaten all over the world providing about 8% of the total human energy intake. Potatoes, yams, sugar beet and cassava (or manioc) are all important root crops.

1.4.1 Vegetative propagation

Many plants undergo **vegetative propagation** as a means of reproduction. This means the plant adopts reproductive strategies which rely on modifications to either the root, stem or the shoot (bud) systems, and which do not involve sexual reproduction. Often plants which reproduce vegetatively also have perennating organs which allow the plant to store food to help it survive harsh climatic conditions like a cold winter for example. But this is not always the case as plants which possess perennating organs can be **biennial** or **perennial** which means the plant survives seasonal climatic changes to live for two years or more.

Let us consider some of the vegetative reproduction strategies that some plants have adopted.

(i) Tillering

Tillers are shoots which develop from the base of the plant. Nearly all grass species produce tillers of one form or another and although these grasses will also produce flowers and undergo sexual

reproduction, tillering is often very important as a method of asexual reproduction.

Tillers generally arise during the vegetative phase of growth that happens before flowering. They develop from **axillary buds**, which are situated at the nodes in the axil of each leaf. So tillers are technically branches, but because the meristem or growing region of the plant is at the base of the stem in grass species, appear to grow out from the base of the plant instead of the stem. This is shown in Figure 11.

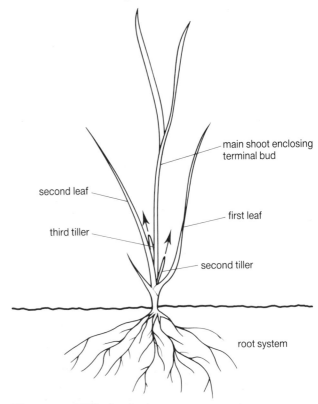

main shoot enclosing terminal bud

second leaf

first leaf

third tiller

second tiller

root system

Figure 11   Tillering in grasses

Each tiller develops its own **adventitious root** system and is, in theory, quite capable of an independent existence as a separate, new plant but there are connections between each tiller so materials are exchanged between them. Tillers grow rapidly during their vegetative phase and are also capable of producing roots so tillering is a means of vegetative reproduction which allows very rapid spread of the plant.

(ii) Runners

The **runners** formed by species such as strawberry plants (*Fragaria*), are like normal stems in structure but they grow horizontally above the ground. Runners are characteristic of rosette forming species whose flat leaves radiate out from a short stem, like the strawberry, and are a vegetative means of establishing new daughter plants away from the parent.

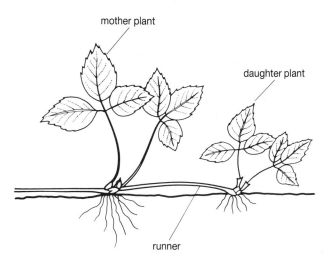

Figure 12   Strawberry runners

This type of vegetative propagation involves a parent plant developing a number of runners, which themselves develop daughter plants at intervals along their length. Each runner may extend for some distance, and give rise to a vast number of new plants. This extraordinary capacity for reproduction has been exploited by strawberry growers, to clone large numbers of identical plants from parent plants selected for particularly favourable characteristics, such as fruit size or disease resistance.

(iii) Corms

Corms are short, swollen vertical underground stems found in species like the crocus. They consist of a swollen stem base surrounded by **scale leaves** which are the remains of the previous season's leaves. Unlike a bulb there are no fleshy leaves. At the end of the growing season, new corms are pulled underground by **contractile roots**. Corms usually contain one or more bulbs.

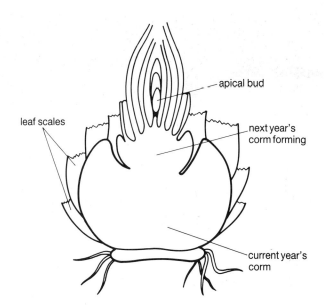

Figure 13   Crocus corms

(iv) Tubers

The potato plant produces runners which have swollen tips. These tips then develop into food storage tubers which are agriculturally very important as a relatively cheap and easily produced source of carbohydrate food. When enough earth is heaped up around the base of the stem, one potato plant will send out numerous runners (or rhizomes). Each one of these grows horizontally below the soil surface. As the season progresses, the tips of the runners begin to swell with stored food, mainly starch, produced by photosynthesis in the leaves. The epidermis of the runner around the swollen tip is replaced by a thin layer of corky material for protection. This is the potato tuber, the part of the plant which is eaten. New plants develop from auxillary buds (or 'eyes') found in a spiral arrangement on the tuber.

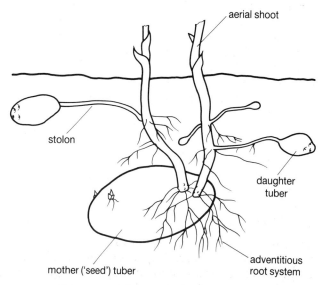

Figure 14   Potato tuber

(v) Bulbs

True bulbs, like onion (*Allium*), are modified shoots constructed very much like the buds on a typical plant. They consist of a very short disc-shaped stem called the **basal plate**, which is made up of a series of nodes, stacked closely together, one on top of the other. Scale leaves are formed which protect the fleshier, swollen, food-storage structures. The bulk of the bulb tissue is formed from the swollen bases of last season's leaves:

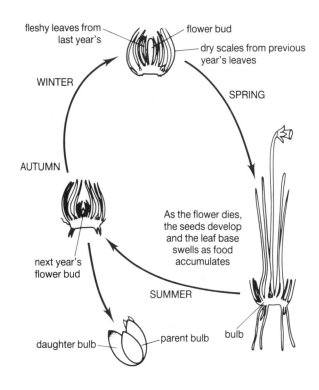

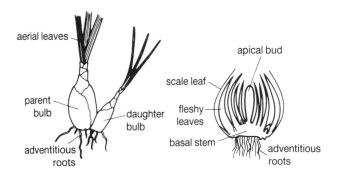

Figure 15   Bulb formation and structure

Each year the apical growing point develops. As it does so, it draws on the food reserves in the scale leaves of the bulb below ground, to form photosynthetic leaves and a flowering stem. During the growing season, once leaves have formed, carbohydrates produced by photosynthesis will be passed back down the plant to be stored in the bulb.

QUESTIONS

1   The following table compares the food sources from which protein and energy are obtained by the human populations in India and in the USA.

Sources of protein and energy in the diet

FOOD GROUP	PERCENTAGE OF TOTAL PROTEIN		PERCENTAGE OF TOTAL ENERGY	
	INDIA	USA	INDIA	USA
Energy rich foods				
Starchy food (cereals, vegetable)	67	24	72	28
Sugary foods (honey, chocolate, sugar)	–	–	9	17
Fats and oils (ghee, butter, margarines)	–	–	4	17
Protein rich foods				
Meat, poultry, eggs and fish	<2	48	0.5	24
Milk and meat products (sausages)	9	24	5	11
Seeds and pulses (lentils, nuts, beans)	20	2	9	3

a)   Draw pie charts to compare the food sources which provide energy in the diets of Americans and of Indians.

b)   Compare the sources from which the two populations obtain their protein. How do you explain the differences you observe?

c)   'By studying these figures, it could be said that the diet consumed in India is nutritionally more healthy than that consumed in the USA'. You have read this statement, alongside the table above, in a magazine. Write a letter to the editor of the magazine explaining why you agree or disagree with the claims.

2   Look carefully at the graph.
a)   From the graph what effect does low temperature storage appear to have on apple seed?
b)   When do you think would be the best time to sow apple seed?

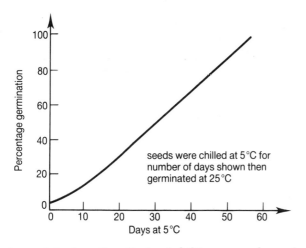

A graph to show the effects of chilling on apple seeds

3 a) You are a wheat farmer looking for new seed from your supplier. What properties would you want the wheat seed you plant to possess, and why?

b) One way of testing seed viability is to measure the percentage germintion. Why do you think the percentage germination of a seed type measured in the laboratory does not necessarily correspond to a high rate of emergence of the same seed type when it is planted in the field?

c) Why might you want to plant seed with *low* vigour? If you wanted to sow seed with a low vigour, what extra precautions would you take to make sure as many of the seeds developed into healthy plants in the field as possible?

d) The germination of weed seeds is intermittent and not synchronised like the germination of crops. Why is this
   (i) an advantage to the weed?
   (ii) a disadvantage to you as a farmer?

4 The table shows the changes in dry mass of the endosperm and entry of maize seeds during germination.

TIME FROM START OF GERMINATION (days)	DRY MASS OF ENDOSPERM (mg/g)	DRY MASS EMBRYO (mg/g)
0	150	1
1	126	2
2	125	3
3	103	3
4	77	10
5	56	15

a) Why is the dry mass of the seeds measured, and not the fresh, wet mass?
b) What is the main disadvantage of measuring dry mass?

c) How might the disadvantage you describe in b) affect the results shown?
d) Use the figures in the table to plot a graph showing changes in the mass of the embryo and the endosperm in maize seeds during germination.
e) What is happening to the endosperm during germination? Why do you think this is happening?
f) What is happening to the embryo during germination? Why do you think this is happening?
g) Describe an experiment to investigate whether oxygen is required for the germination of seeds.

5 In an investigation of germination in barley, a single grain was cut in two so as to separate the endosperm and the embryo. Each part was placed face downwards on starch agar in a petri dish. A second grain was prepared in the same way but placed on starch agar containing gibberellic acid. After two days, each petri dish was flooded with iodine solution. The results are shown in the diagram below.

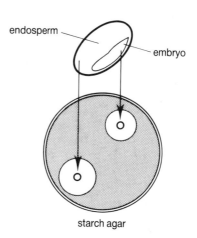

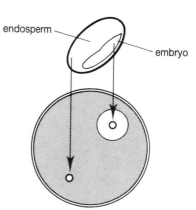

starch + gibberellic acid agar
Colour after flooding with iodine solution:
▓ blue-black   ☐ colourless

Diagram showing the results of the experiment

a) (i) Before placing the cut grains on the agar, they were washed in sterilising solution. Explain the reason for this.

(ii) In setting up this experiment, why do you think it is important not to touch the surface of the agar with your fingers?

b) What information do the results of this investigation suggest about the production of amylase in the endosperm of an intact grain?

(AEB 1992)

6 (i) Describe the structure of an anther and a pollen grain.

(ii) Outline the formation of pollen grains, emphasising how variation is introduced into the process.

(iii) Suggest how the knowledge of pollination mechanisms is important to fruit growers.

(UCLES 1991)

7 Imagine you are a fruit producer in an African country. The buyer for a new fruit wholesalers in Britain wants to import bananas from you but is anxious to make sure they won't be over-ripe when they arrive. Write a letter to explain how you would manage a consignment of bananas to ensure that they do not over ripen whilst they are being transported to Britain.

BIBLIOGRAPHY

Attridge, T.H. (1990) *Light and Plant Responses.* Edward Arnold.

Audus, L.J. (1972) *Plant Growth Substances.* Leonard Hill.

Hill, T.A. (1980) *Endogenous Plant Growth Substances. (IOB).* Edward Arnold.

Janick, J. (1986) *Horticultural Science.* W.H. Freeman.

Langer, R.H.M. *How Grasses Grow. (IOB).* Edward Arnold.

Villiers, T.A. *Dormancy and the Survival of Plants. (IOB).* Edward Arnold.

Whatley, J.M., Whatley, F.R. (1980) *Light and Plant Life. (IOB).* Edward Arnold.

CROP GROWTH AND PRODUCTIVITY

LEARNING OUTCOMES

After studying this chapter you should be able to:
- define the terms 'growth' and 'development' as applied to crop plants,
- evaluate and discuss the advantages of a range of parameters which can be used to determine growth,
- construct and interpret growth curves,
- calculate and compare absolute and relative growth,
- distinguish between 'productivity' and 'harvestable yield'.

2.1 CROP GROWTH AND DEVELOPMENT

Producers of commercial crops aim to encourage growth and development to produce maximum crop yield. The yield or productivity of a crop is affected by several environmental factors which we will discuss in later chapters. The farmer or grower must have an understanding of how these factors affect the growth of his crop in order that the environment can be manipulated or controlled to obtain maximum productivity.

Growth and development of crop plants involves many different physiological and biochemical processes. Plant growth results in at least one of the following:
- an irreversible increase in dry mass
- an irreversible increase in cell number
- an irreversible increase in cell size.

Plant development is often defined as the progress of a plant towards maturity, involving an increase in cell and tissue complexity. Figure 16 shows a Zadoks key which is used to describe growth and development in wheat. This key charts different stages of growth in detail, providing over 90 points of reference. This sort of key is a useful reference for farmers as the effectiveness of many agrochemicals increases if they are applied at certain stages of growth.

Notes 1) Spring cereals five leaves unfolded include only leaves on main shoot.
2) Winter wheat pseudostem erect when leaf sheath exceeds 5 cm (2 ins).
3) Nodes seen or felt on main stem after peeling back leaf sheaths.

Growth stage description	First leaf unfolded	Two leaves unfolded	Three leaves unfolded main shoot and 1 tiller	Four leaves unfolded main shoot and 2 tillers	Five leaves unfolded main shoot and 3 tillers (Note 1)
Decimal code (Zadoks)	11	12	13–21	14–22	15–23

Seedling growth · *Tillering*

Leaf sheaths (pseudostem) erect (Note 2)	First node detectable (Note 3)	Second node detectable (Note 3)	Flag leaf just visable	Flag leaf ligule/collar just visible	Roots swollen
30	31	32	32	39	45

Stem elongation · *Rooting*

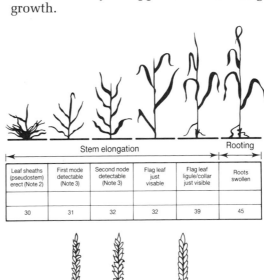

Growth stage description	First awns visable (barley)	First spikelet of ear just visible (wheat)	Emergence of ear (inflorescence) complete
Decimal code (Zadoks)	49	51	59

Ear emergence

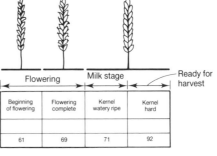

Beginning of flowering	Flowering complete	Kernel watery ripe	Kernel hard
61	69	71	92

Flowering · *Milk stage* · Ready for harvest

Figure 16 The Zadoks growth stage key for wheat

2.1.1 Measuring growth

Crop growth can be determined using a number of different measurements including length, area, volume and mass. The method chosen will depend upon the plant and the reason it is being measured, but the most common parameter used is **mass**. Two values for mass can be obtained: **fresh** or wet mass, and **dry** mass.

2.1.2 Measuring fresh mass

Fresh mass has the advantage that it is easy to measure and is a non-destructive process which can be repeated over a period of time. If the growth of an individual plant is being monitored it can be determined simply by weighing, although the term fresh mass can be misleading because plants absorb varying amounts of water. This means that the water content of any two plants of the same species and age can be very different and this will be reflected in their fresh mass. This is illustrated in Figure 17.

2.1.3 Measuring dry mass

Measuring the dry mass of a plant reflects true growth more closely as all the water is removed. This is done by incinerating the plant in an oven at 110 °C and recording the weight at intervals until no further change in mass is observed. As the measurement of dry mass is a destructive process it is unsuitable for use in some experiments such as continuous monitoring of growth.

Figure 17   The disadvantage of using fresh mass to determine growth

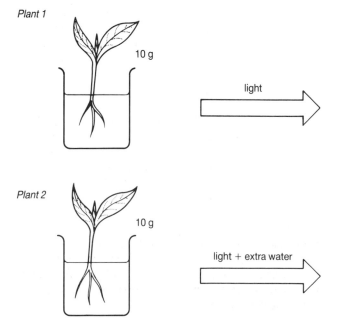

2.2 GROWTH CURVES

If plant growth is measured over a period of time and the data plotted onto a graph, a curve is produced. Plant growth curves usually follows a typical **sigmoid**, or S-shaped, pattern. This is illustrated in Figure 18.

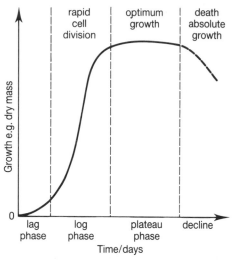

Figure 18   A sample growth curve

As you can see from the graph, the following phases of growth can be distinguished:
1 The **lag phase** – a period of little growth.
2 The **log phase** – a period where the rate of growth is at its maximum. The plant grows **exponentially** so the amount of growth is always proportional to the amount of material or number of cells already present.
3 The **decelerating phase** – a stage of development where the rate of growth begins to slow down.
4 The **plateau phase** – overall growth ceases.

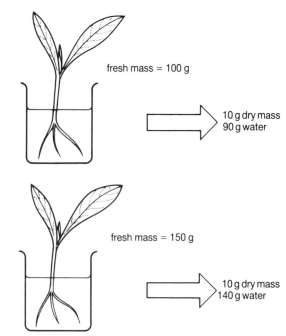

2.2.1 Absolute growth rate

To find out how much a plant has increased in size the absolute growth rate (G) can be calculated using the following equation:

$$G = \frac{M_2 - M_1}{T_2 - T_1}$$

M_1 = mass at time T_1
M_2 = mass at time T_2

The absolute growth rate (G) of a plant is its mean increase in dry mass per unit time. However, while it gives an indication of how much the plant has increased in size, it does not tell us how fast the plant is growing. This is because it does not take into account the size of the plant at the start of the growth period measured and so can only be used to compare the rate of growth in identical plants of the same species and age.

2.2.2 Relative growth rate

Another way to calculate plant growth is to measure the relative increase in growth. Relative growth rate (RGR) is a measure of the rate at which dry matter increases per unit of dry matter present, with time. The relative growth rate can be used by farmers to compare the rate of growth in a crop at different times of its life cycle, or to compare the growth of different crops.

It is important that growers and farmers monitor the rate at which their crops are growing and developing in order to make the most efficient use of the agrochemicals they apply, and that they can determine the best time to harvest so that maximum yield is obtained.

$$RGR = \frac{\log_e M_2 - \log_e M_1}{T_2 - T_1}$$

M_1 = mass at time T_1
M_2 = mass at time T_2

Now try Investigation 4 Measuring Growth Rate in the *Plant Science in Action Investigation Pack*.

2.3 PRODUCTIVITY AND YIELD

2.3.1 What do we mean by productivity?

The amount of food available to higher trophic levels in an ecosystem is determined by the efficiency with which a plant community traps chemical energy from the sun and converts it into useful products. These products are generally termed **dry matter** as they represent the organic material produced by the plant when the water within the plant is removed. The amount of water within a plant can vary considerably and so the measurement of dry matter is a more accurate way of assessing the availability of organic products. Irrespective of their growth habit, the net productivity of all plants depends on the twin processes of photosynthesis and respiration. The rate at which the products of photosynthesis accumulate in the plant is called the **net assimilation rate** or the **net primary productivity**.

2.3.2 Net assimilation rate

Net assimilation is measured in grammes of dry matter per unit area per day ($gm^{-2}day^{-1}$). The accumulation of chemical energy allows the plant to produce new cells and for the existing ones to grow and thus to produce dry matter. Net assimilation rate, therefore, gives us an accurate measure of plant growth which can be used to compare plants at different stages of their life cycle.

2.3.2 Harvestable yield

Most of our arable crops are harvested for their final weight of grain and the contribution made by their leaves and stems is often unimportant. Net assimilation provides the necessary biomass for storage in leaves, stems, roots, flowers and grain. For most arable crops it is the *reproductive* fraction of total productivity that constitutes their economic yield. In other crops, economic yield depends on the *vegetative* fraction of growth: sugar for example, is extracted from the stems of sugar-cane and from the roots of sugar beet. Yield is a measure of the 'harvestable' portion of the crop, and is expressed in tonnes per hectare (t/Ha).

The final productivity of a crop depends on the processes of development and growth that do, or do not occur, between the germination and the harvesting of seeds. Ultimately, both development and growth are controlled by the interaction of genes and environment.

Only parts of the whole plant are used. This constitutes the harvestable yield.

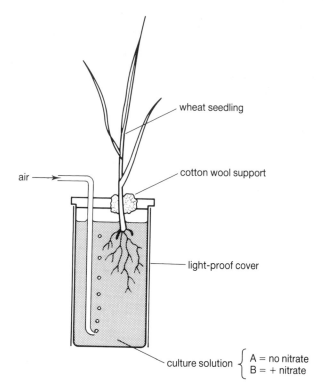

Wheat seedlings grown in different mineral cultures

b) Figure 16 shows the Zadok's Growth Stage key for wheat. Use the key to identify the following stages of growth:

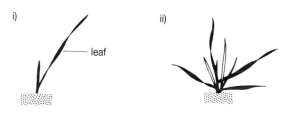

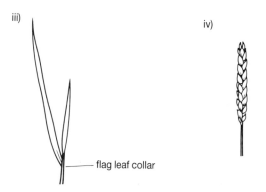

QUESTIONS

1 You will remember that when measuring growth a number of parameters can be used. The most common is mass. This can be measured in two ways – fresh weight or dry mass. Imagine you wished to compare the growth of wheat seedlings grown in two different mineral cultures for a week.

a) Describe how you would measure the fresh mass of the seedlings.
b) What are the advantages and disadvantages of measuring fresh mass?
c) Explain how you would measure the dry mass of the seedlings.
d) What are the advantages and disadvantages of this method of determining mass?

2 a) Define the following
        (i) plant growth
        (ii) plant development.

c) Why do you think this type of key is useful to farmers and growers?

3 a) Plant productivity or yield depends on photosynthesis and respiration. Why is the productivity of a crop equal to the net productivity, not the gross?

b) What is meant by the term **biomass**?

c) What environmental factors do you think will affect values of net photosynthesis?

4 In an ecological investigation to determine the net primary productivity of a field of wheat, random samples of wheat plants were collected from a field. The dry mass of each of these samples was determined and the energy content of the dry wheat measured experimentally. From the figures collected, the net primary production in kJm^{-2} was calculated.

a) Describe how you would have,
(i) collected random samples of wheat from the field,
(ii) found the dry mass of one of these samples.

b) The apparatus shown was used to measure the energy content of a sample of the dry wheat:

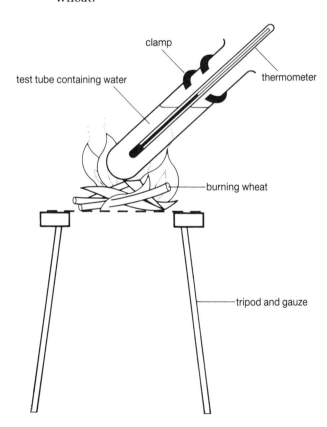

clamp

test tube containing water

thermometer

burning wheat

tripod and gauze

Measuring the energy content of wheat

(i) List the measurements that it would have been necessary to take in order to make the necessary calculations.
(ii) Describe two important sources of error that arise when using this apparatus.

(AEB 1992)

5 The data in the Table below gives information about the dry mass of broad bean seedlings 10 and 12 weeks after planting.

TIME (weeks)	DRY MASS (g)
10	40
12	55

a) Calculate the absolute growth rate for this sample.

b) Calculate the relative growth rate for this sample.

c) Why is the relative growth rate more accurate than the absolute growth rate?

6 a) Explain how photosynthesis and respiration are related to the accumulation of dry matter in a plant.

b) What do you understand by the term 'harvestable yield'?

c) What part of the plant makes up the harvestable yield in the following crops:
(i) potato crop
(ii) barley crop
(iii) carrot crop
(iv) apple crop
(v) strawberry crop?

7 Net assimilation rate (NAR) and leaf area index (LAI) are two ways of measuring crop performance.

a) (i) State how the LAI is measured for a field crop.
(ii) Comment on the relevance of the LAI to an understanding of crop productivity.

b) (i) State how the NAR is calculated for a field crop.
(ii) What information about crop productivity is gained by calculating the NAR at different times during the growing season?

The diagram below shows the results of an investigation of the leaf area index (LAI) and net assimilation rate (NAR) for four field crops, carried out at the Rothamsted Experimental Station in the 1940s.

c) With reference to the graphs:
(i) Comment on the growth of spring cereals.
(ii) State the advantage of growing a winter cereal rather than a spring cereal.
(iii) Suggest two reasons why it is not advisable to plant crops such as potatoes early in the year.

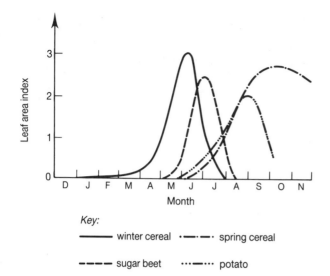

Key:

——— winter cereal   ·—·—· spring cereal

– – – sugar beet   ······ potato

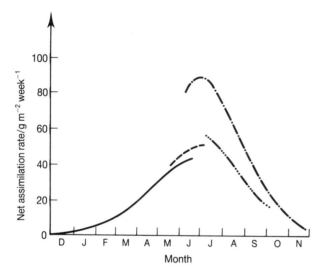

Graphs showing the leaf area index and net assimilation rate of four common crops

d) Crop growth rate (CGR) is an index of productivity. It is calculated as:

$$CGR = NAR \times LAI$$

(i) From the results shown in the graphs, calculate the maximum CGR for sugar beet. Show your workings.

(ii) Suggest two ways in which farmers can increase the CGR of a field crop.

(UCLES 1992)

BIBLIOGRAPHY

Leopold, A.C., Kriedemann, P.E. (1975) *Plant Growth and Development*. McGraw-Hill, New York.

Sestak, Z., Catsky, J., Jarvis, P.G. (1971) *Plant Photosynthetic Production* (chapter 10). Dr. W. Junk.

3 CROP PRODUCTIVITY AND LIGHT

LEARNING OUTCOMES
After studying this chapter you should be able to:
- list the plants' metabolic processes which require light,
- describe the differences between the mechanisms of photosynthesis used in C_3 and C_4 plants,
- describe the fate of the solar energy which reaches the surface of a leaf,
- calculate 'leaf area index' and explain the effect of this on crop growth,
- define 'leaf area duration' and explain its importance to the growing crop,
- distinguish between long day, short day and day neutral plants,
- discuss the role played by light in the control of flowering.

3.1 THE SUPPLY OF SOLAR RADIATION

The Sun provides the Earth with a constant supply of radiation. Part of this radiation is discernable as visible light. The human eye is able to distinguish the different colours of light, each of which has a different wavelength. Figure 19 shows the light spectrum.

Plants utilise the visible wavelengths of light for several metabolic processes for example:
- Dry matter production by photosynthesis. (Productivity).
- To control flowering in some species by photoperiodism.
- To break seed dormancy in some species (see Chapter 1).

3.2 THE EFFECTS OF LIGHT ON CROP PRODUCTIVITY

3.2.1 Mechanisms of photosynthesis

We know that within the major arable species, there are two photosynthetic mechanisms. In some plants, the initial fixation product is a three-carbon compound. These are known as C_3 plants. In these crops, some of the initially fixed CO_2 is lost through photorespiration, and therefore these plants often have a lower net rate of assimilation.

Figure 19   Visible light is only part of the radiation received from the Sun.

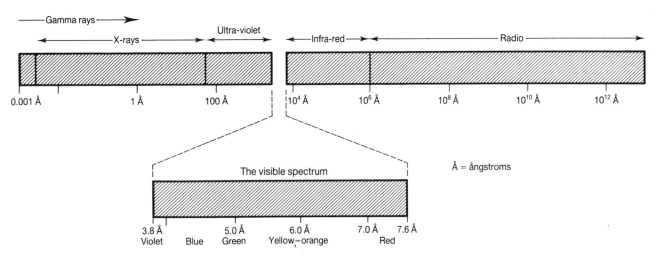

21

In other plants, a four-carbon compound is the initial fixation product so they are called C_4 plants. Here, most of the respired CO_2 is *refixed* and there is little, or no, loss of carbon by photorespiration. So, when C_3 and C_4 plants are grown in similar atmospheric and soil environments, C_4 species typically produce dry matter more rapidly than C_3 species (although they require more solar energy to do so). The C_3 species include all the temperate crops like barley and wheat, whereas C_4 crops include most tropical cereals and grasses, such as maize, sorghum and sugar-cane. Table 3.1. shows the effects of photorespiration on productivity in tobacco. While respiration losses accounts for about a 13% reduction of productivity, photorespiration accounts for a further loss of 55%.

Table 3.1 The effects of photorespiration on the productivity of tobacco

	CARBON DIOXIDE FIXED $g/cm^2/hr.$
Gross productivity	4.7
Respiration loss	0.6
Photorespiration loss	2.6
Net productivity	1.5
Net productivity when photorespiration is blocked	4.1

(Zelitch, 1975 *Photosynthesis, Photorespiration and Plant Productivity* Academic Press)

Irrespective of species, the net rate of assimilation of photosynthetic products is the difference between the rate at which CO_2 is converted (to carbohydrate) and the rate at which carbohydrate and other compounds are respired (to provide the energy necessary for metabolic processes) [Monteith, 1981]. The pathway for the net movement of CO_2 into a leaf is schematically illustrated in Figure 20.

3.2.2 Light interception

Environmental factors which affect a plant in such a way as to limit growth and development are known as **limiting factors**. Under conditions of no limiting factors, such as no disease, sufficient water etc, net photosynthesis depends on the plants' ability to intercept solar radiation (light). Figure 21 shows the typical fate of 100 units of light energy that reaches the leaf of a crop plant from the Sun.

As you can see from the diagram only 50% of the incident light hitting the leaf is a suitable

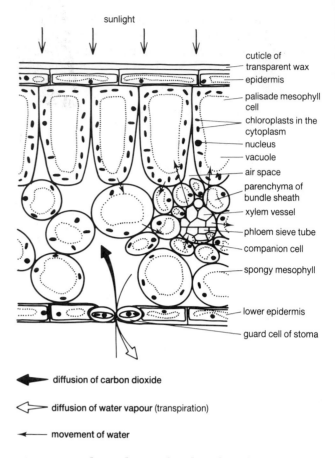

diffusion of carbon dioxide

diffusion of water vapour (transpiration)

movement of water

Figure 20 The pathway of carbon dioxide through the leaf

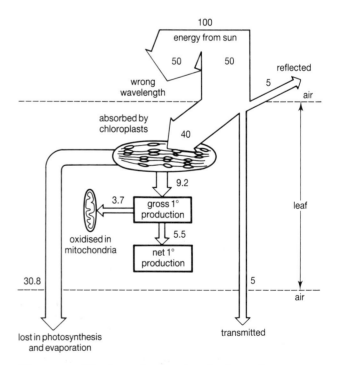

Figure 21 The fate of 100 units of solar radiation reaching the leaf of a crop (Hall 1979)

wavelength for photosynthesis (that is, light in the red and blue ranges). Of this 5% is reflected by the leaf, and 5% is transmitted straight through the leaf, leaving 40% to be absorbed by the chloroplasts. In reality only 9.2% of the initial solar energy is used to produce dry matter (gross primary production), of which 5.5% will find its way into the net primary products (dry matter to provide material for growth). The outstanding amount of energy is used to fuel metabolic processes within the plant, or lost by photorespiration.

3.2.3 Leaf area index

In order to photosynthesise and produce dry matter a crop must intercept solar radiation and absorb carbon dioxide. As this process takes place in the leaves, the size and lifetime of its foliage determines the rate at which the crop can accumulate dry matter, and thus grow. A useful indicator of the size of a crop canopy and hence its potential productivity is its **leaf area index** (L) which is the ratio of leaf area to ground area covered by the canopy, taking only one side of each leaf into account. The leaf area index provides a useful indication of the potential productivity of a standing crop as it is the measure of the area of leaf available for the interception of light.

The leaf area index of a crop can be expressed as

$$L = N_P \times N_S \times A_S$$

where
N_P is the number of plants per unit of ground area
N_S is the number of leaves per plant
A_S is the mean area of leaves per plant.

(Squire 1990)

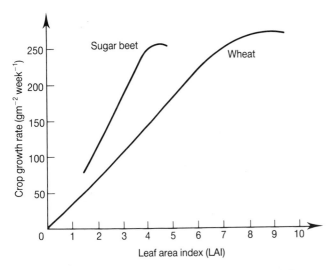

Figure 22   The relationship between crop growth rate and leaf area index in wheat and sugar beet

When the leaf area index of a crop is one, the total leaf area of the crop is equal to the area of the ground it is growing on. For example if a standing crop of wheat growing in a one metre square of land had a leaf area index of one, this would mean that the total area of the top surface of all the crop leaves would be one metre squared.

Wheat reaches its maximum growth rate when its leaf area index is about seven. This means that the crop's leaves cover seven times the area of the ground the crop is growing on. As the leaves on a wheat plant are narrow and erect, it must produce as many leaves as it can providing a large surface area to ensure that it intercepts the light it needs to produce dry matter.

Sugar beet, on the other hand has broad, flat leaves. This means that it will reach its maximum growth rate at a lower leaf area index, as this type of leaf is able to absorb incident solar radiation more efficiently.

The narrow erect leaves of wheat

The broad flat leaves of sugar beet

> **Now try Investigation 5 Measuring Leaf Area Index in the *Plant Science in Action Investigation Pack.***

3.2.4 Leaf area duration

Leaf area duration (LAD) tells us the timespan over which a crop has a particular leaf area. It gives a measure of the opportunity that a crop has to produce dry matter. The longer the leaves have been out the greater the opportunity for the crop to trap sunlight, and the higher its productivity or yield.

3.2.5 Light and the accumulation of dry matter

It has been shown experimentally, that the relationship between dry matter accumulation (net assimilation) and the interception of solar radiation is approximately linear for many crops, at least during the vegetative stage of growth. Thus, the rate of dry matter production per day can be expressed as the amount of light intercepted by the crop in 24 hours and the efficiency with which it uses that light in the production of dry matter.

Hence the linear relationship shows that, in principle, uniform crops produce dry matter at a rate which is almost proportional to the amount of radiant energy that they intercept. This has been shown to be reasonably constant for a number of species. However, actual dry matter production often falls short of potential dry matter production as shown in Table 3.2.

Table 3.2   Comparison of potential and actual dry matter accumulation (yields) in the UK (1978)

PRODUCTION	WHEAT	SUGAR BEET	POTATOES	BEANS
Average Farm (t/Ha)	5.00	6.80	32.00	2.80
Best Farm (t/Ha)	10–12	9–10	60–80	5–6
Potential (t/Ha)	12–15	12–15	90–95	9–10

3.2.6 Photosynthesis and the food chain

Plants are able to utilise solar radiation to produce food from carbon dioxide and water. This means that they form the first **trophic** level of all food chains. A food chain charts the way in which energy and food is transferred from organism to organism. Some of the dry matter produced by a plant is used by the plant as a substrate for respiration and is therefore lost energy.

The rest is available as food for herbivores and omnivores. Herbivorous organisms form the second tropic level of the food chain. They metabolise some of the food eaten, storing some as muscle to be passed on as meat for organisms at the next level. Food and energy will be passed on in this way until they reach a consumer at the top of the food chain, for example, a human being. But,

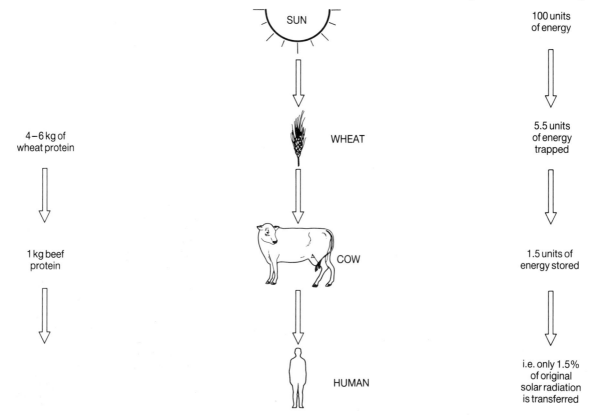

Figure 23   The approximate fate of energy and protein within a simple food chain

at each level of the food chain about 90% of the energy transferred is lost through respiration and to decomposing organisms. Only 10% is available as food for the next level. Therefore, you can see that it is more economical to eat organisms from the lower trophic levels as shown in Figure 23.

3.3 LIGHT AND THE CONTROL OF FLOWERING

Productivity in many species is related directly to the formation of flowers. The plant's reproductive organs are contained in flowers which are also responsible for both seed and fruit production. Flower production may contribute to the harvestable yield of a crop plant in a number of ways:

- the **flowers** themselves may be harvested, for display, as pot plants and cut flowers or for food, for example cauliflowers
- the **seeds** they produce may be harvested, for example, cereal grains, peas, beans
- the **fruits** they produce may be harvested, for example, apples, strawberries, tomatoes.

It is therefore important that the grower has an understanding of how flowering is initiated and controlled within the plant.

3.3.1 The life cycle of flowering plants

Study the life cycle of the flowering plant as summarised in Figure 24.

As you can see, providing the right environmental conditions exist, flowering will occur in the **mature phase** of plant development. Plants vary a great deal in the rate at which they proceed from the juvenile to the mature phase. In some cases it may happen abruptly, in others it takes place slowly and may gradually occur over a period of many years. There is also variation in the extent of the transition. In some species, such as wheat and sunflower, transition from the juvenile or vegetative phase to the mature or flowering phase is complete. This means that all the shoots on the plant switch from vegetative growth to flower production. In other species, such as runner beans, peas and tomatoes, vegetative growth and flowering can occur at the same time.

3.3.2 Controlling the time of flowering

In the wild, the time of flowering is closely related to the environment. An ability to match a life cycle to a seasonal climate is clearly important if the flowers are to survive the climatic extremes of

Figure 24 The life cycle of the flowering plant

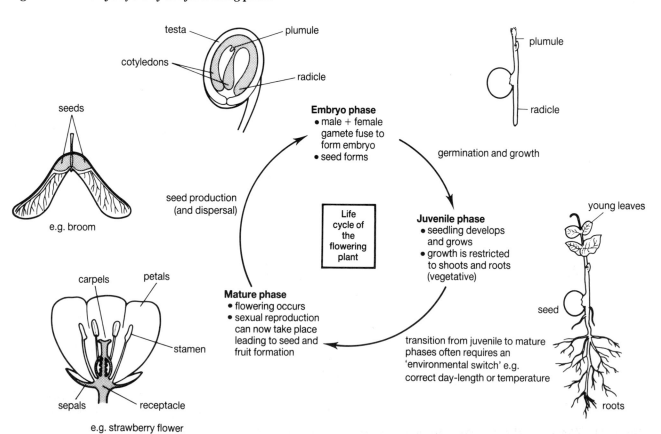

winter cold or summer drought. Flowering is generally initiated by the 'switching on' of certain genes, so that the plant changes from leaf production to flower production. This is a slow and irreversible process and plants have to get it right first time if they are to synchronise their flowering with the season and with others of the species. In many species the switching on of the genes happens automatically when the plant is of a certain age. But in other species an environmental signal is needed to trigger flowering. This signal is most commonly either **day-length** or **temperature**, although the nutrients available in the soil and the action of plant growth regulators (or hormones) within the plant may also have an influence. The effects of temperature on flowering will be discussed in Chapter 4.

3.3.3 The effects of day-length

Most plants flower at a particular time of year. In climatic zones where winter carries the risk of serious frost damage (and where temperatures fall below 0 °C for long periods) plants need to monitor the process of the seasons. Plant activities must be synchronised with the seasons, so that frost sensitive stages of the life cycle are passed through during periods when the temperature is warmer. Flowering is a particularly vulnerable activity as pollen and ovules are very susceptible to frost damage. In fact, the most reliable environmental variable to monitor, is not temperature but day-length – or more correctly night length – which, unlike temperature, does not vary from year to year, (see Table 3.3).

Table 3.3 December day-length and average temperatures at Sutton Bonington, Leicestershire.

YEAR	AVERAGE DAY-LENGTH (hrs/mins.)	AVERAGE TEMPERATURE (°C)
1973	7.53	4.8
1974	7.53	7.8
1975	7.53	5.3
1976	7.53	1.7
1977	7.53	5.6
1978	7.53	3.5

In general, plants can be divided into three groups on the basis of their flowering response to day-length.

Day neutral plants can flower at any time of the year, irrespective of day length. If conditions are suitable for growth and the plant has reached the mature phase, then flowering can occur. Many weed species such as dandelion, groundsel and annual meadow grass fall into this group, as do some glasshouse crops such as cucumber and tomato and many species from tropical and sub-tropical regions. The majority of these species are either wind pollinated or, if they are insect pollinated, then they are able to pollinate themselves so that seed can still be produced, even if the conditions are unsuitable for insects.

Short day plants flower when the day-length is less than 12 hours. In the UK this is the autumn. Plants such as rice, and cotton fall into this group, as do chrysanthemums which will be discussed later.

Long day plants flower in the spring and summer, in the UK, when the day-length is longer than 12 hours. This group includes many of the biennial crops and winter-sown cereals and also spring sown plants such as clover, spinach and radish.

3.3.4 Photoperiodism

Species of plants which are sensitive to day-length are called **photoperiodically controlled** species. The duration of both darkness and light as well as light intensity and wavelength, all effect the plant's response. Some species such as *Xanthium pennsylvanium*, will respond to exposure to only one suitable light cycle. Other species like *Chrysanthemum moriflorium* require several cycles of ever decreasing day-length.

There are two types of response to day-length: **obligate** or **facultative**. Obligate photoperiodic plants havde a well developed **critical** day-length, above or below which flowering will not occur. In facultative photoperiodic plants, flowering will occur sooner if a suitable day-length is experienced.

3.3.5 Day-length and the phytochrome system

The length of the night (dark period) is measured by the phytochrome system. This is discussed in detail in Chapter 1. You will remember that the phytochrome is a blue-green pigment which occurs in two interchangeable forms of isomer: the P_R (cis-isomer) absorbs red light (665 nm) and the P_{FR} (trans-isomer) absorbs far-red or infra-red light (725 nm). When P_R absorbs red light, it rapidly changes into P_{FR} and when P_{FR} absorbs far-red light, it rapidly changes into P_R. P_{FR} will also slowly change into P_R in the dark.

In the species in which this system operates, the relative proportions of the two isomers of phytochrome are important in the initiation of flowering. Short day plants require low levels of P_{FR} before they will flower, so require short periods of light and longer periods of dark. During the dark period, P_{FR} will be slowly converted into P_R thus causing levels of P_{FR} to remain relatively low. Long day plants, however, require high levels of P_{FR} to initiate flowering and so long periods of light are required, during which the conversion of P_{FR} to P_R is prevented.

It is thought that the phytochrome system controls flowering by stimulating or inhibiting the production of a hormone. This hormone has not yet been isolated, but has been called **florigen**.

Now try Investigation 6 Comparing Long and Short Day Plants in the *Plant Science in Action Investigation Pack.*

QUESTIONS

1 a) One way to improve productivity is to develop varieties of C_3 species, like wheat, which mature earlier in the growing season before the hot summer weather occurs. How do you think this will overcome the problems of photorespiration?
   b) What do you see as the disadvantages associated with growing this type of variety?
   c) Suggest another way in which a breeding programme could remove the problems of photorespiration.

2 a) To attain a crop growth rate of $200 \, \mathrm{gm}^{-2} \, \mathrm{wk}^{-1}$, sugar beet only requires a leaf area index of 3.0, whereas wheat requires one of 7.0. Why?
   b) Why is it an advantage to achieve optimum leaf area index as early as possible in the year?
   c) Imagine you are a sugar beet farmer. How would you ensure that optimum leaf area is reached as early as possible in the year? What factors might prevent you from being successful?
   d) Why might plants with narrow, erect leaves such as grass, have an advantage over broad leaf plants, like clover late in the growing season?

3 Examine the graph opposite which shows the absorption spectra from phytochrome. It shows that phytochrome is made up of two isomers.
   a) Which of the isomers absorbs redlight?
   b) Which of the isomers absorbs far-red light?
   c) The two isomers are interchangeable. Which of the two isomers would you expect to measure at a high level in a plant that is in the dark?

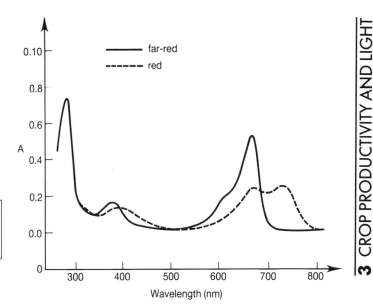

The absorption spectra for phytochrome

   d) The relative levels of the two isomers of phytochrome control flowering in some species. Explain how flowering is controlled by phytochrome in
      (i) short day plants.
      (ii) long day plants.

4 The diagram below shows the effects of different lighting treatments on a short day species and a long day species.

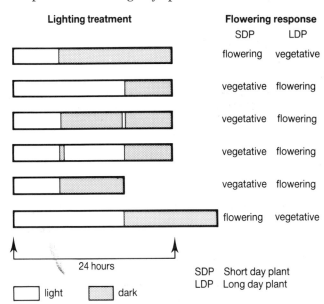

The effect of duration of dark period on flowering.

   a) Explain why short day plants do not flower when exposed to
      (i) light treatment 3?
      (ii) light treatment 4?
   b) Explain why long day plants flower when exposed to
      (i) light treatment 3?
      (ii) light treatment 4?

c) Chrysanthemums are short day plants. Imagine you are a commercial grower. How would you manage your crop in order to supply flowering plants all year round?

5 The table below shows the number of cuttings taken from *Hedera helix* which successfully rooted. Cuttings were taken from juvenile and mature wood.

AGE OF WOOD	PERCENTAGE OF CUTTINGS TAKING ROOT
Juvenile	100
Mature	17

a) Taking cuttings from a plant is a method of asexual reproduction (vegetative reproduction). Examine the data in the table. What does the data suggest about the effects of maturity in *Hedera helix* on the plants' ability to reproduce asexually?
b) What physiological change is likely to accompany maturity?
c) Suggest why horticulturalists may wish to manipulate the onset of maturity in
  (i) tomatoes (*Lycopersicon sp*)?
  (ii) chrysanthemums (*Chrysanthemum morifolium*)?

6 The table below shows the yield of protein which can be obtianed from one hectare of arable land.

LAND USE	YIELD OF PROTEIN (kg)
Growing crops	400 kg plant protein
Raising cattle	25 kg animal protein

a) Explain the differences in protein yield shown in the table.
b) Use this information and other data which you can collect to prepare a short talk either in favour of or against vegetarianism. This talk could be prepared as an individual or in a small group. Set up a debate with other members of the class. Try to present your ideas clearly, using posters, and other visual aids as available.

BIBLIOGRAPHY

Attridge, T.H. (1990) *Light and Plant Responses.* Edward Arnold.

Chapman, S.R., Carter, L.P. (1976) *Crop Production.* W.H. Freeman.

Chrispeels, M.J., Sadava, D. (1977) *Plants, Food and People.* W.H. Freeman.

Hall, D.O., Rao, K.K. (1987) *Photosynthesis.* Edward Arnold.

Kendrick, R.E. *Phytochrome and Plant Growth.* Edward Arnold.

Luckwell, L.C. (1981) *Plant Growth Regulators in Crop Production.* Edward Arnold.

Zelitch, I. (1975) *Photosynthesis, Photorespiration and Plant Productivity.* Academic Press.

TEMPERATURE AND CROP GROWTH

4.1 TEMPERATURE AND CROP PRODUCTION

Soil and air temperatures are very important to crop production. Most plants are very sensitive to even the smallest changes in temperature at most stages in their growth and development. Plants can therefore be classified as either **cool** or **warm** season plants. Cool season plants will grow at temperatures as low as 4 °C, whereas warm season plants will not grow at temperatures below about 10 °C. In some species temperature provides the critical environmental stimulus which induces flowering or seed germination. These processes are often initiated after a period of cool treatment called **vernalisation**.

4.1.2 Soil temperature

The temperature of the soil depends mainly on the amount of solar radiation which reaches the soil surface, and the efficiency with which it is then absorbed. If soil is covered with vegetation, little sunlight is able to penetrate the crop canopy. Dark soils absorb heat more rapidly than ones

which are light in colour. Soils with a high water content absorb the solar radiation well but because water has a high specific heat capacity (it requires 4.2 Joules of energy to raise the temperature of one gram by 1 °C), these soils do not warm up very quickly.

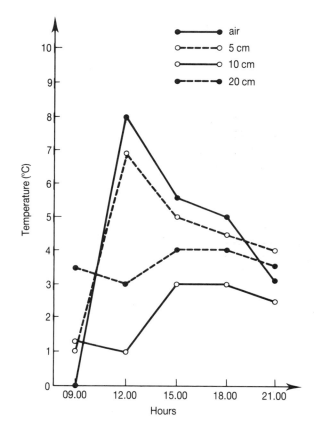

Figure 25   Temperature variation in air and soil

The temperature of the soil will naturally vary with the time of day, and depth penetrated as shown in Figure 25. You can see from the graph that the temperature of the soil near the surface is closely related to the air temperature. The temperature of soil from deeper levels shows less variation. The graph also shows how both air and soil temperature peaks at around midday, when solar radiation is also at a peak.

Soil temperature will also vary with season. Figure 26 shows how the soil temperature varies according to the time of year in Nigeria.

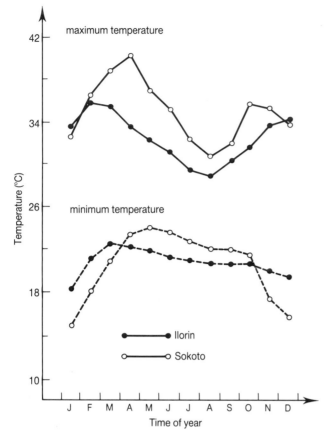

Figure 26   Annual variation in soil temperature in Sokoto, Nigeria

The temperature of the soil will affect the rate at which water evaporates from the soil surface. It will also have an effect on the rate at which soil organisms are able to break down organic matter and the rate at which seeds germinate. In general, the higher the soil temperature, the more rapidly these processes occur.

4.1.3 Air temperature

Air temperature also varies with season and region of the world as well as with the time of day. The amount of solar radiation available will determine the temperature of the air and so average air temperatures tend to decrease as we move away from the equator as shown in Figure 27.

The air temperature varies throughout the day according to daily or a **diurnal** pattern with the temperature peaking at midday in a similar way to the soil temperature.

4.1.4 Temperature and crop development

Plants are susceptible to changes in temperatures at various stages in their life cycles. Temperature affects plant growth by controlling the rate at which biochemical and physiological reactions occur. This is because these reactions are controlled by enzymes which are very temperature sensitive. Changes in temperature of only a few degrees can have quite major effects on crop growth and productivity. The effects of temperature changes on germination of seed, grain production, leaf production and root elongation in millet are shown in Figure 28.

Figure 27   Map to show worldwide average air temperatures (°C) in July

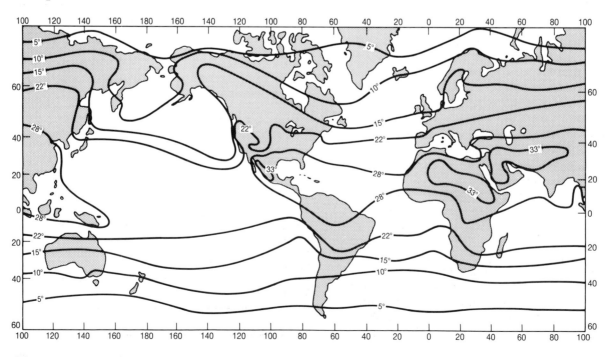

a)

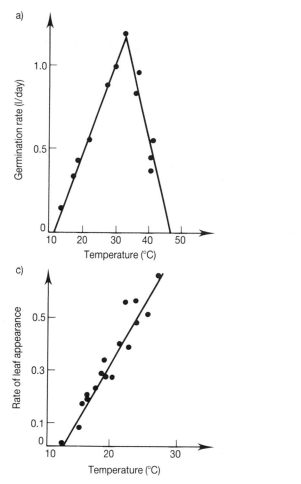

b)

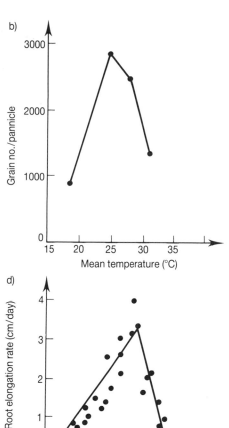

c)

d)

Figure 28   Graphs to show the temperature effect on growth and development in millet

4.2 CARDINAL TEMPERATURES

As with all living things, temperature controls the rate at which biochemical reactions can occur. At different stages of development, each crop species will have three key or **cardinal** temperatures.

(i) Minimum temperature at which biochemical reactions can occur.
(ii) Maximum temperature at which biochemical reactions can occur.
(iii) Optimum temperature at which biochemical reactions occur at a maximum rate.

Cardinal temperatures will determine the geographical areas in which crop species can be successfully grown. For a crop to grow and develop successfully, it usually requires a certain number of days at its optimum temperature. The length or duration of the growing season available is as important as the temperature received by the plant. The duration of the growing season can be calculated in **degree-days**. A degree-day is used to measure the temperature requirement in a plant. For example, maize requires 3–4 months (120 days) at about 30 °C for optimum development.

That is, it requires 3600 degree-days (120 × 30 = 3600).

Tables 4.1 (a) and (b) show the cardinal temperatures for several crops at different stages in their life cycle.

Table 4.1 (a) Cardinal temperatures for germination

	TEMPERATURE (°C)		
CROP	MINIMUM	OPTIMUM	MAXIMUM
Maize	10	32–35	44
Rice	12	30–32	38
Sugar-cane	21	32–37	46
Soyabean	10	25–30	40
Cotton	13	25–32	39
Groundnut	12	24–30	38
Barley	3	15–22	30
Potato	12	20–28	30
Wheat	5	20–24	32

Table 4.1 (b) Cardinal temperatures for vegetative growth

CROP	TEMPERATURE (°C)		
	MINIMUM	OPTIMUM	MAXIMUM
Maize	10	28–32	38
Rice	22	28–30	40
Sugar-cane	13	26–29	41
Soyabean	12	25–32	35
Cotton	13	24–30	40
Groundnut	15	28	38
Barley	6	16–18	40
Potato	8	20–25	32
Wheat	5	18–24	41

As you can see the crops shown in Tables 4.1(a) and 4.1(b) can obviously be divided into those which will grow in temperate climates like barley and those which grow only in hotter areas such as sugar-cane. Temperatures required for germination relate to soil temperature and show the temperature that is required after vernalisation has occurred (see Chapter 1). It is interesting to note that temperatures required for *growth* relate to air temperatures.

4.3 EXTREME TEMPERATURE DAMAGE

We sometimes hear in the news, stories about crops damaged by extremes of temperature. High temperatures will cause scorching and eventually damage the enzyme systems within the plant causing death. High temperatures are usually associated wth lack of water. In hot climates water will evaporate more rapidly from the soil and from the leaves of the plants growing there. This means that as less water is available to the plant, seeds will fail to germinate and more established seedlings will fail to thrive and so eventually die. In tropical areas the problems associated with high temperatures and water availability are made worse because rainfall is not evenly distributed throughout the year. In these regions rain tends to fall over a period of weeks or months, known as the **rainy season**, and then no more rain falls throughout the year. This means that if no rain falls during the rainy season it will be a full year before it is likely to rain again. Hence drought conditions can develop very rapidly making it extremely difficult to produce food crops. In many tropical regions, for example Sudan and Ethiopia, this pattern of little or no rainfall during the rainy season and extremely high daytime temperatures, has led to the total failure of crops and the situations of famine, which are becoming common.

An emergency feeding centre at a refugee camp, Sudan.

Low temperatures will slow down enzyme reaction rates and therefore, the growth of a plant. Excessively low temperatures can lead to frost damage. As a plant is cooled to 0 °C it will wilt due to the formation of callous plugs in the vascular bundles and cytoplasmic membranes begin to break down, leak their contents and the leaves will turn brown. Exposure to temperatures below 0 °C causes further damage.

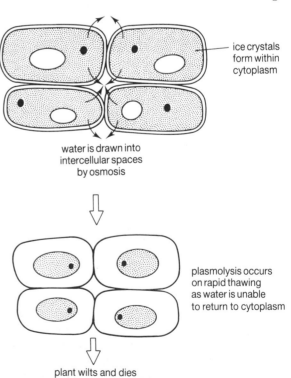

ice crystals form within cytoplasm

water is drawn into intercellular spaces by osmosis

plasmolysis occurs on rapid thawing as water is unable to return to cytoplasm

plant wilts and dies

Figure 29   Damage to plant cells by extremely low temperatures

The high water content of the vascular bundles and the cytoplasm causes the formation of ice crystals. Water is drawn out of the cell into the intracellular spaces by osmosis. If the plant is thawed rapidly, this water does not return causing the cells to collapse and the plant to die. Plants can be 'hardened' to the effects of frost by exposing them to low temperatures at an early developmental stage. Plants which are first hardened have cytoplasm with a lower freezing point and their cell membranes have increased permeability.

The effect of extremes of temperature can have serious commercial implications. In Florida, orange groves can be destroyed by severe frosts.

4.4 THE EFFECTS OF TEMPERATURE ON FLOWERING

As we discussed in Chapter 3, some species of plants require an environmental signal to initiate flowering once they have reached maturity. This signal may be light or temperature. In most cases, where temperature induces flowering, it is low temperatures, between −1 and 10 °C, which are effective and in nature this normally means that the plant has to experience a cold winter before it can flower. A plant's requirement for low temperature is called **vernalisation** and is characteristic of plants which have a life cycle which spans two years (a biennial life cycle). These plants generally grow vegetatively during the first year and then flower during the second year.

As a general rule, plants which will flower in response to vernalisation, fall into two groups: **obligates** and **facultates**. For obligates vernalisation is essential for flowering. While facultates will flower without receiving a cold treatment, but they will flower much sooner if vernalised.

In most species, vernalisation does not directly trigger flowering; it prepares the plant so that it is able to produce flowers on exposure to high temperatures and longer days during the following spring.

How vernalisation works is not absolutely clear but it is thought that low temperatures alter the balance between two growth regulators or hormones which naturally occur within the plant. It is thought that cold temperature removes the hormone abscissic acid which leaves room for the gibberellins, a group of hormones, to predominate. Some biennials, such as carrots and lettuce which generally require vernalisation, may be stimulated to flower, at normal

temperatures by applying gibberellic acid. We also know that in many of the cases where vernalisation is necessary, day-length is also important. As most biennials require long days to flower they tend to do so during May and June.

Biennial spring flowers, such as primroses will not flower unless they are vernalised.

Some perennials (plants which flow each year) also show a need for a low temperature trigger before flowering. Many common spring flowers such as celandines, primroses and violets are known to require exposure to cold conditions each winter in order to flower in the spring.

4.4.1 Winter and spring cereals

The first clue to the importance of temperature as a regulator of flowering, came from work on cereals, carried out during the first quarter of the twentieth century. There are varieties of cultivated cereals available, especially wheat and barley, which should be sown either in autumn or

Winter cereal varieties require vernalisation.

in spring. The autumn sown varieties (winter varieties), have a facultative cold requirement and produce higher yield if they are subjected to low temperatures after sowing. Winter wheat is usually sown in September or October so that it overwinters as small plants and flowers during the following spring. Spring sown varieties do not have a requirement for cold and so grow up and flower in one season. Yield of spring varieties is not affected by temperature.

4.4.2 Sugar beet

Sugar beet (*Beta vulgaris*) is grown extensively in the eastern counties of the UK. It is naturally a biennial plant, which produces vegetative growth durings its first year, and flowers during its second year. Sugar beet is grown for the enlarged tap root which it produces underground during its first year. This is rich in sucrose, which can be extracted and used to produce granulated sugar and other products. The size of the tap root and its sugar content will depend on the amount of energy the leaves of the plant are able to trap by photosynthesis during its first growing season.

The enlarged tap root of the sugar beet

Sugar beet has an obligate cold requirement and is therefore susceptible to vernalisation. Vernalisation of the growing plant will cause it to flower, thus starting the mature, reproductive phase of its life cycle. This takes some of the food stored in the tap root and reduces the sugar content. As the plants are grown for their sugar content, farmers avoid vernalisation by harvesting the mature beet before they can be chilled by winter temperatures. Plants which begin to grow flowers are called bolters and these are not useful to the grower because of the reduced sugar content of the beet. Modern sugar beet varieties produce very few bolters. As they appear so rarely

they can be pulled up by hand. This prevents seed dispersal which must be avoided if the farmer is growing crops in rotation. Commercial seed suppliers will encourage bolting so that seed can be produced for sale to farmers.

Table 4.2 shows the effects of vernalisation on several varieties of sugar beet. If seed is sown in the spring, the incidence of bolting is relatively low. If, however, the seed is sown in the autumn, the plants are subjected to chilling over the winter and are therefore likely to bolt.

Table 4.2   Numbers of bolters in sugar beet per 1000 at different sowing dates

VARIETY	BOLTERS PER 1000	
	Normal sowing (April)	Early sowing (Early March)
Amethyst	2	20
Bravo	1	26
Ragina	1	13
Salohill	2	24

(Data based on figures from NIAB, Cambridge Variety Trials 1986.)

Bolting in sugar beet

Mature sugar beet plants require many hours of sunlight in order to build up dry matter in the large tap root. Sunlight is trapped by the leaves during photosynthesis and by this process, sugar is produced and stored in the root. So it is important that seed is sown early enough for plants to produce maximum leaf area by the summer when maximum sunlight is available. The grower, therefore, must balance the risk of vernalisation and the need to maximise leaf area as quickly as possible when selecting dates for sowing and harvesting his crop.

> **Now try Investigation 7 Inducing Flowering in Vegetables by Vernalisation in the *Plant Science in Action Investigation Pack*.**

QUESTIONS

1 For a crop to grow successfully it requires a certain number of 'degree days'.

CROP	OPTIMUM GROWING SEASON (DAY)	OPTIMUM TEMPERATURE (°C)
Maize	120	30
Rice	145	21
Groundnut	125	21
Soyabean	120	21
Cotton	190	34
Tobacco	105	24

(a) Calculate the number of degree days required by the crops shown in the Table.

(b) Explain why these crops are unsuitable for growth in temperate climates.

2 The Table below shows the optimum cardinal temperature for grain development in several species.

CROP	OPTIMUM CARDINAL TEMPERATURE FOR GRAIN DEVELOPMENT (°C)
Barley	18–20
Cotton	22–25
Groundnut	25–30
Maize	25–28
Millet	25–28
Potato	18–22
Rice	30–32
Sugar-cane	24
Tobacco	18–22
Wheat	18–24

a) What do you understand by the term optimal cardinal temperature?

b) How is the optimal cardinal temperature of a species related to the minimum and maximum cardinal temperatures for that species?

c) Which of the crops listed appear to be suitable for growth in temperate areas, for example the UK?

d) Explain why some of the crops which have suitable optimal cardinal temperatures for growth in temperate regions do not survive in these areas?

3 a) How do you think enzyme systems within a plant would be affected by temperature damage?

b) Explain, using diagrams, what happens to a plant cell when it is exposed to frost and then thawed rapidly.

c) How does 'hardening' prevent this damage?

4 The Table below shows the various wheat varieties sown at different dates.

	SEPT.	OCT.	NOV.	DEC.	JAN.	MAR.
Norman	9.9	10.7	10.0	7.7	6.9	6.2
Wembley	8.3	9.3	9.3	7.6	6.8	7.2
Gawain	9.9	10.2	9.6	8.6	7.8	6.6
Avalon	8.4	9.5	9.3	7.6	6.1	5.8

YIELD (t/Ha) sown

(Data taken from 'Wheat – A guide to Varieties' Plant Breeding International, Cambridge.)

a) Use the data shown in the table to draw a line graph for each wheat variety. Use the same axes for all varieties.

b) From your graph what is the general relationship shown between sowing date and yield?

c) For each variety state the optimum sowing date.

d) Winter wheats have a **facultative cold requirement**. What does this mean?

e) Suggest two reasons why sowing winter wheat in the early autumn may increase yields.

f) Spring wheats do not have a facultative cold requirement. Which of the varieties shown is most likely to be a spring variety?

g) The weather in Scotland is often cold and wet during the autumn, making it unsuitable for sowing wheat. Why might it be a advantage to the farmer to buy spring wheat seed for planting instead of a winter variety?

h) What appears to be the disadvantage of spring wheats?

i) Suggest why certain species have evolved with this requirement for chilling?

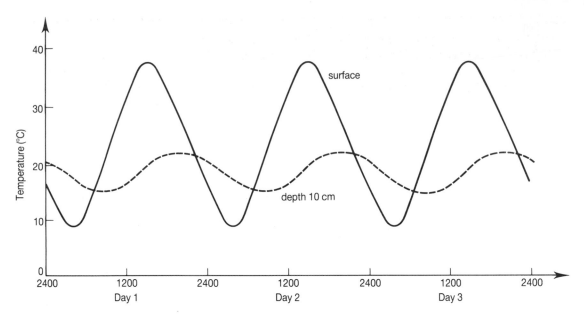

Graph showing how the temperature of the soil fluctuates

5 The diagram above shows the temperature fluctuations at the surface and at a depth of 10 cm in bare soil on three successive, clear, sunny days.
   a) Explain the shape of the curve for surface temperature.
   b) State two differences between the curve for temperature at a depth of 10 cm and that for the surface. Explain these differences.
   c) How will the presence of vegetation affect the daily fluctuations of temperature?
   d) Describe one way in which soil temperature influences plant growth.

(JMB 1990)

BIBLIOGRAPHY

Forbes, J.C., Watson, R.D. (1992) *Plants in Agriculture.* CUP.

Onazi, O.C. (Editor) (1988) *Crop Science and Production in Warm Climates.* Macmillan Publishers Ltd.

Sutcliffe *Plants and Temperature. (IOB).* Edward Arnold.

SOIL SCIENCE AND CROP PRODUCTION

After studying this chapter you should be able to:
- describe the processes which lead to the formation of soil,
- explain the importance of humus in soil, and interpret a soil profile,
- classify soils according to particle size and relate this to the overall structure and composition of the soil,
- describe cultivation techniques used when preparing a seed bed,
- list the reasons why a plant requires water,
- define the terms 'field capacity' and 'permanent wilting percentage' and use these to calculate water availability,
- explain the terms 'evapotranspiration' and 'transpiration ratio',
- describe the effects of too little water on crops and evaluate the use of irrigation on a selection of species grown in different environments,
- describe the effects of too much water on crops and evaluate the use of soil drainage techniques,
- list the main nutrients required by the plant for healthy growth,
- discuss the response of crops to the application of nitrogen fertilisers,
- compare the use of inorganic and organic fertilisers and evaluate the merits of each,
- discuss the potential risks to human health and the environment of increasing the use of nitrogen fertilisers.

5.1 CROP PRODUCTIVITY AND SOIL TYPE

The productivity of a crop plant is affected by the type of soil in which it is grown. Plants obtain water and nutrients from the soil in the form of minerals. The capabilities of a soil to provide nutrients will largely depend on the parent rock from which it was formed. The soil's physical structure and chemical composition is also important when assessing which crops can grow where.

5.2 SOIL FORMATION

As you can see from Figure 30, soil is formed by the physical and chemical breakdown of rock over an extended period of time. Soil formation involves three main stages:

(i) Chemical and physical **weathering** of the parent rock to produce smaller mineral particles.
(ii) The chemical modification of some of these particles.
(iii) The addition and decay of organic material.

5.2.1. Weathering

Weathering is the term used to describe the breakdown of rock into smaller particles. Rock can be weathered both physically and chemically.

Table 5.1   Weathering of rock

Physical weathering	• extremes of temperature can cause rock to fracture as a result of being heated and cooled,
	• rain water falling into crevices in rocks can become frozen causing the water to expand and causing rocks to crack,
	• running water e.g. rivers and streams, and wind cause rock to become eroded causing small rock particles to be washed or blown away
	• roots of plants colonising the rock grow through cracks and enlarge them.
Chemical weathering	• rocks are dissolved by chemicals in rain water (acid rain) e.g. carboxylic and sulphuric acids, and sometimes nitric acid.

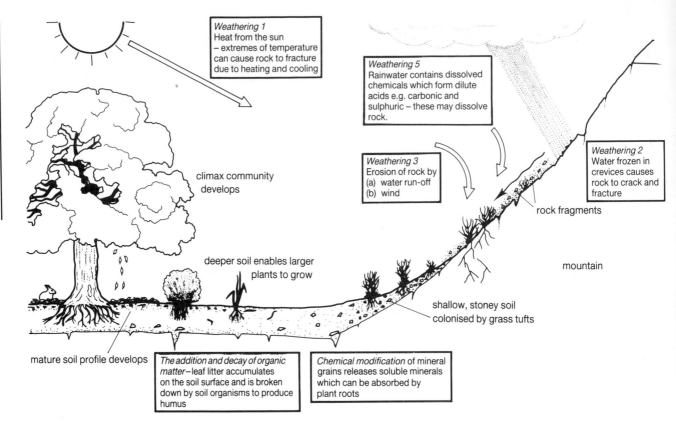

Weathering 1
Heat from the sun – extremes of temperature can cause rock to fracture due to heating and cooling

Weathering 5
Rainwater contains dissolved chemicals which form dilute acids e.g. carbonic and sulphuric – these may dissolve rock.

Weathering 3
Erosion of rock by
(a) water run-off
(b) wind

Weathering 2
Water frozen in crevices causes rock to crack and fracture

climax community develops

rock fragments

mountain

deeper soil enables larger plants to grow

shallow, stoney soil colonised by grass tufts

mature soil profile develops

The addition and decay of organic matter – leaf litter accumulates on the soil surface and is broken down by soil organisms to produce humus

Chemical modification of mineral grains releases soluble minerals which can be absorbed by plant roots

Figure 30 Soil formation

5.2.2. The chemical modification of soil particles

Weathering is the process whereby grains of mineral particles are released from the rock. The comparison of these grains will be determined by the type of rock from which they were formed, however, the most common components include oxygen, silicon, aluminium and iron. Some of these particles can then be chemically modified and broken down so that soluble minerals are released. For example, the milky white mineral feldspar ($KAlSi_3O_8$) reacts with carboxylic acid in rain water to form soluble potassium carbonate and kaolinite. It is important that mineral particles are broken down further so they are available as soluble compounds and can be absorbed by the roots of plants.

5.2.3. The addition and decay of organic material

Soil contains varying amounts of organic material which is derived from the organisms which live within the soil or on the soil surface. Leaf 'litter' accumulates on the soil surface where it is broken down by decomposers to form **detritus**.

Earthworms and other soil organisms help mix this into the soil gradually, and as it decomposes, it forms a brown colloidal substance called **humus**. A high humus content encourages plant growth because:

- humus provides a food source for micro-organisms living in the soil. They release nitrogen, and other elements essential to plant growth, as a by product of their metabolism,
- humus retains moisture so soil is less likely to dry out,
- humus binds soil particles together to produce a crumbly soil which allows roots to grow through it more easily and which is well aerated,
- soils with high humus contents tend to absorb heat more rapidly and so processes such as germination can occur more quickly.

The amount of humus within the soil will depend on the rate at which leaf litter is being added to the soil, the rate at which it is being broken down and the rate at which the nutrients released are being used up by the plants growing there. The rate of decay of leaf litter is influenced by:

- oxygen content – most decay organisms require oxygen for respiration and so decay tends to be more rapid in well aerated soils,
- acidity – naturally occurring decay organisms do not thrive in acidic soils so the decay of

organic material in acidic soils tends to occur slowly as seen in pine forests,

- temperature – high temperatures, for example in the tropics, speed up the growth of decay organisms and so speed up decay (see Table 5.2).

Table 5.2 Time taken for the decay of organic matter produced on one hectare of land

ENVIRONMENT	TIME TO DECAY YEARS
Tropical rainforest	1.7
Temperate deciduous forest	4.0
Conifer forest	14.0
Tundra (arctic)	50.0

5.2.4 The maturation of soil

The rate at which soil is formed will depend on a number of factors. Soft rocks, such as sandstone, will break down more rapidly than hard rocks, like granite. High rainfall will speed up the process of weathering, especially if the rain which is falling carries corrosive pollutants, such as sulphur dioxide, as will high temperatures. If rainfall is high, there is an increase in the amount of water which drains through the soil. This **percolating** water causes soluble minerals and small particles to be leached out of the top layers of soil and redistributed lower down the water table. This results in the formation of a mature soil which is made up of at least three layers or **horizons** as shown in Figure 31 which shows a typical soil profile.

- Horizon A is only 20–30 cm thick and is commonly known as topsoil. It is rich in minerals as organic material is converted into humus in this layer. The roots of most plants only penetrate this layer. If rainfall is high the percolating water can leach soluble minerals from the topsoil. This is more of a problem in some types of soil than others as we will see later.
- Horizon B, the subsoil, lies beneath the topsoil. Water and minerals which have been leached out of the topsoil build up in the subsoil. Farmers often plough deep enough to break up the subsoil and so release extra nutrients for their crops.
- Horizon C is the parent rock from which the soil has formed as a result of weathering.

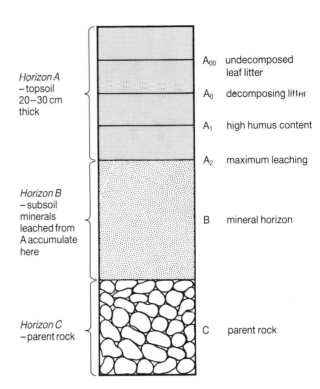

Figure 31   An example of a profile of a mature soil showing layers or horizons

5.3 SOIL STRUCTURE

There are three main types of soil particle which can be formed as a result of weathering rocks. They are classified according to their particle size as shown in Table 5.3. Figure 32 shows the comparative size of each of these particles.

Table 5.3   Classification of soil particles (International Society of Soil Science)

PARTICLE TYPE	DIAMETER (mm)
Sand	2–0.02
Silt	0.02–0.002
Clay	<0.002

coarse sand

2–0.02 mm in diameter

fine sand

silt

0.02–0.002 mm in diameter

clay

<0.002 mm in diameter

Figure 32   The comparative size of soil particles

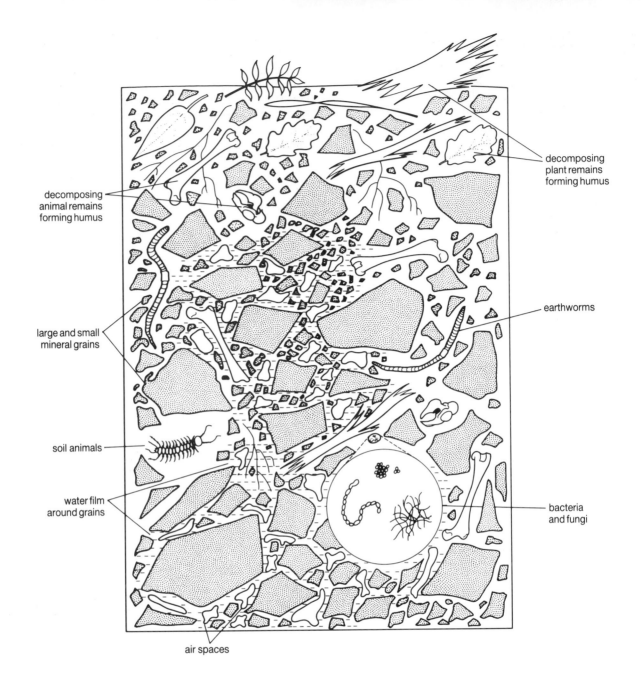

decomposing
plant remains
forming humus

decomposing
animal remains
forming humus

earthworms

large and small
mineral grains

soil animals

water film
around grains

bacteria
and fungi

air spaces

Figure 33   Soil structure

The particles within a soil are naturally clumped together so that there are pores between each particle. These pores, or capillaries, are filled with air. When water percolates through the soil some is trapped within these air spaces forming a film on the surface of the soil particles. A fertile soil will contain both of these as roots need both water and oxygen.

Soil generally contains a mixture of these components shown in Figure 33. The proportion of each type of particle that is present in the soil will determine the soil's properties. Table 5.4 shows the composition of three common types of soil.

Table 5.4   Composition of common soil types

SOIL TYPE	SAND %	SILT %	CLAY %	WATER HOLDING CAPACITY
Sandy loam	85	6	9	Poor – leaching of water and nutrients possible.
Loam	59	21	20	Good.
Clay	10	22	68	Too good – waterlogging a problem.

5.3.1 Sandy loam soils

A sandy loam has a high sand content – in the region of 85%. Sand particles are relatively large (0.02–2 mm in diameter) and so they are unable to clump closely together. This means that large pores or capillaries exist between the particles as shown in Figure 34. These capillaries are filled with air so sandy loam soils are well aerated allowing soil organisms and plant roots to thrive. The large capillaries also allow water to percolate through the soil rapidly. Sandy loam soils drain well and so are easy to plough as they do not become waterlogged. However, as water drains through the soil it leaches minerals from the topsoil and deposits them in the subsoil layer where they are unavailable to plant roots. The nutrients may become available to crops in following seasons if the land is ploughed. A soil with a high sand content tends, therefore to be less fertile.

a) The arrangement of particles in sandy soil

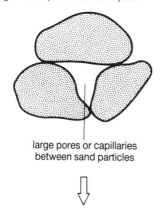

large pores or capillaries
between sand particles

b) Structure of a sandy soil

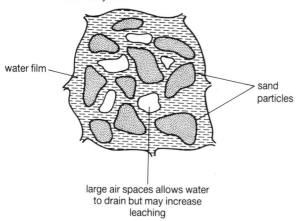

water film

sand
particles

large air spaces allows water
to drain but may increase
leaching

Figure 34   Sandy loam soils contain a large proportion of sand particles.

5.3.2. Clay soils

Clay soils are a lot harder to cultivate than sandy soils because they contain a high proportion of clay particles. Clay particles are very small (less than 0.002 mm in diameter) and so are able to clump very closely together (as shown in Figure 35). This means that the capillaries formed between the particles are very small and so the soil tends to have a low air content. The small capillary size also means that clay soils do not drain well and so although the leaching of minerals is less of a problem, clay soils tend to become waterlogged and so are difficult to plough or dig. When a soil becomes waterlogged, the capillaries fill with water leaving little space for air. This means that aerobic soil organisms and plant roots are unable to thrive unless the soil is cultivated.

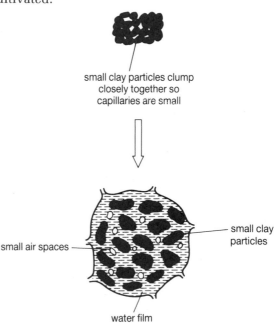

small clay particles clump
closely together so
capillaries are small

small air spaces

small clay
particles

water film

Figure 35   Clay soils contain a high proportion of clay particles.

5.4   PREPARING THE SOIL FOR CROP PRODUCTION

Before crops can be sown in a field the soil must be **cultivated** so that a fertile seed bed is produced. This is usually done mechanically and involves a series of procedures.

5.4.1 Primary cultivation: ploughing and subsoiling

Ploughing is often the first stage in the preparation of the seed bed. The plough blade slices into the topsoil at a depth of about 25 cm. Behind the blade lies a mould board which lifts the slice of soil and leaves it on the side of the furrow produced. Ploughing has several purposes. It helps break-up the soil down so that aeration can take place. This may also help water to drain more easily because larger air spaces exist between soil particles. This is especially important in heavy clay soils. Ploughing can add to soil fertility by mixing organic material from the surface into the soil. If the soil is ploughed deeply enough, nutrients which have leached into the subsoil may also be mixed into the topsoil layer and so become available for crops. Ploughing also helps remove weed seedlings by uprooting them from the soil.

Ploughing

Subsoiling is an alternative to ploughing. During this process the soil is loosened to a similar depth without inversion. This action improves soil drainage and so is particularly suitable for use on soils with a high content of clay.

5.4.2. Secondary cultivation: harrowing, discing and rolling

Once the soil has been ploughed it can be further cultivated using a number of techniques. The processes which are selected by the farmer at this stage will depend on the crop to be sown, soil type and weather conditions. These processes will break down the soil further so that the topsoil is suitable for seed germination.

5.4.3 Minimum cultivation

One of the main functions of soil cultivaton is to remove weeds from the soil before seeds are sown so that there is no competition between the crop and weeds. With the development of herbicides the need for this has declined and many farmers now practice minimum cultivation. This may involve ploughing of the top layers of soil only or using techniques such as direct drilling, where seed is planted directly into the soil without cultivation. Minimum cultivation techniques reduce the amount of time required to prepare the seed bed, therefore reducing the amount of time the soil remains bare and susceptible to erosion. It also reduces labour and fuel costs. However these techniques require high levels of herbicide application.

5.5 WATER AND CROP PRODUCTION

Water is required by the plant for many different processes:
- to obtain minerals from the soil required for healthy growth,
- as a substrate for the process of photosynthesis,
- as an essential medium for the movement of the products of photosynthesis in the leaves to other parts of the plant,
- for transpiration which helps to regulate the internal temperature of the plant by dissipating heat produced during other metabolic processes,
- in non-woody plants to keep the cells turgid and so support the stem,
- protoplasm has a high water content and water can make up over 80% of the wet weight of some crops (see Table 5.5).

Table 5.5   Water content of various crop plants (% fresh weight)

CROP	WATER CONTENT (%)
Carrot	88
Asparagus	88
Cabbage	86
Lettuce	95
Tomato	94
Watermelon	92
Maize	11
Barley	10
Groundnut	5

5.5.1 Rainfall

The amount of rain which falls on an area will depend mainly on two factors:
1 the location of the land,
2 the season.

In general, annual rainfall increases as we move towards the equator to the tropics. However rainfall in tropical areas tends to be confined to a few months of the year when rainfall is persistent and heavy. This rainy season usually lasts for several months, during which time the planting and growth of crops is impossible. Figure 36 shows duration and rainfall during the rainy season in an area of India.

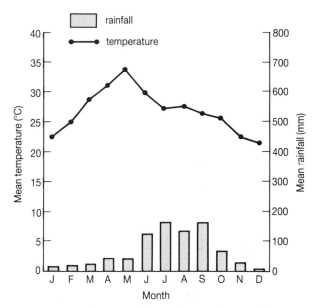

Figure 36   The annual distribution of rainfall in Hyderabad India 1990

Hyderabad is in the centre of the Indian land-mass, fairly flat and hot, so rainfall is limited to the monsoon months. In temperate regions, rainfall is more evenly distributed throughout the year as shown in Figure 37.

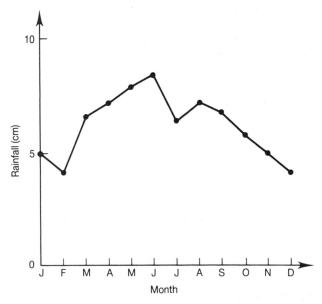

Figure 37   The annual distribution of rainfall in Midwest USA

5.5.2. Water availability

Not all the water that falls as rain is available for plants. Only a small proportion of falling rain is intercepted by vegetation, the rest remains on the surface, where it may evaporate, soak into the soil, or run into waterways. As plants obtain most of their water from the soil through their roots, farmers and growers must ensure that most of the water that falls is able to percolate into the subsoil where it can be stored, and is not lost. This can be encouraged by cultivating the soil so that water can filter through. If soil is too compact or waterlogged, rain water will remain on the surface of the soil.

As we know from section 5.3, the amount of water contained within a unit volume of soil will depend upon the type of particles which the soil contains. A soil with good retaining capacity will obviously provide more water for the plants' growth, but if the retaining capacity is *too* good, waterlogging can occur. The ease with which water can be extracted from a soil by plants can be assessed by measuring two parameters:

- **Field Capacity**: the amount of water held by the soil once it has been left to drain so that all excess water is removed;
- **Permanent Wilting Percentage** (PWP): the amount of water within a soil at which plants growing there become permanently wilted.

The amount of water available for uptake by plants can be calculated by subtracting the permanent wilting capacity from the field capacity.

$$\begin{array}{ccc} \text{Water} & = & \text{Field} & - & \text{Permanent} \\ \text{availability} & & \text{capacity} & & \text{wilting} \\ \text{(mm)} & & \text{(\%)} & & \text{capacity} \\ & & & & \text{(\%)} \end{array}$$

Table 5.6   Field Capacity, permanent wilting percentage and available water for several soil types

SOIL TYPE	FIELD CAPACITY (%)	PERMANENT WILTING (%)	AVAILABLE WATER (%)
Sand	9	2	7
Loam	27	11	16
Clay	39	22	17

> Now try Investigation 8 Comparing the Water Availability in Soils in the *Plant Science In Action Investigation Pack.*

5.5.3 Removing water from the soil – evapotranspiration

You probably already know that water is removed from the soil by two processes:
1 transpiration,
2 evaporation from the soil surface.

Transpiration is the main factor causing water to be absorbed from the soil and moved through the plant. The transpiration-cohesion-tension theory described in 1894 by Dixon and Joly,

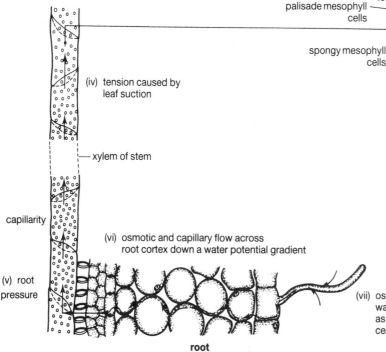

leaf
palisade mesophyll cells
spongy mesophyll cells

(iii) osmotic and capillary flow from leaf xylem to mesophyll down a water potential gradient

(ii) evaporation into air pockets of mesophyll

(i) diffusion into atmosphere (mainly through stomata)

(iv) tension caused by leaf suction

xylem of stem

capillarity

(vi) osmotic and capillary flow across root cortex down a water potential gradient

(v) root pressure

(vii) osmotic uptake by root hair from soil solution down a water potential gradient – water is drawn into the root hairs as it is present in higher concentration in the soil than in the cell sap. This sets up a water potential gradient.

root

Figure 38 Transpiration

suggest that as the leaves of a plant lose water by transpiration, the water vapour in the leaf spaces is replaced by water, ultimately from the soil surrounding the roots. Water moves up the xylem of the stem due to the difference in **water potential** (Ψ) which exists as a result of transpiration. The process of transpiration is discussed in detail in many standard A level Biology texts. It is summarised in Figure 38.

As the Sun warms the soil surface water vapour will be lost by evaporation. The combination of the processes of water loss by transpiration and evaporation is called **evapotranspiration**. For water to be available for crop growth the annual rainfall must exceed annual evapotranspiration. Figure 39 is a graph that shows how rainwater must be stored in the subsoil at certain times of the time if enough water is to be available to overcome evapotranspiration at other times.

Water availability is summarised in Figure 40.

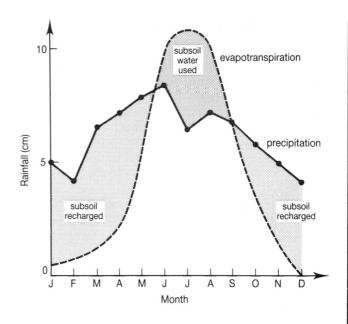

Figure 39   Rainfall must exceed evapotranspiration if water is to be available for crop growth

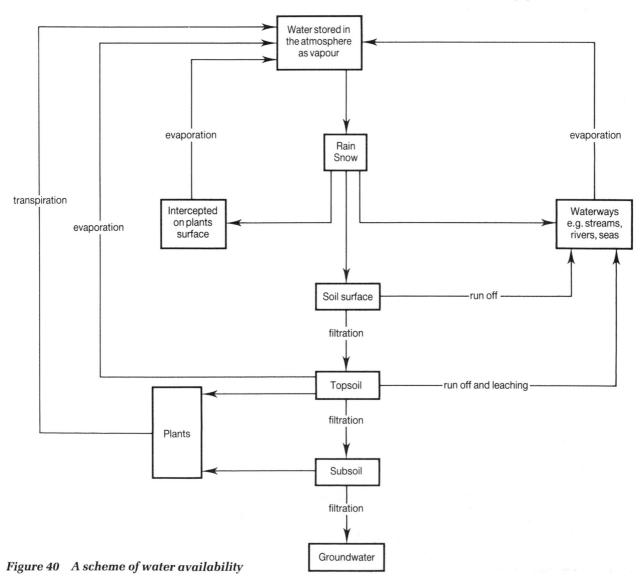

Figure 40   A scheme of water availability

5.5.4 Water use efficiency

Some species of plants are able to use the water available more efficiently than others. Different plants require different amount of water in order to produce dry matter. The **transpiration ratio** describes the amount of matter (grams) that is produced by the plant per kilogram of water absorbed. Plants with high transpiration ratios are unsuitable for growth in dry regions. Some examples of transpiration ratios are shown in Table 5.7.

Table 5.7   Transpiration ratios of some common crops

CROP PLANT	TRANSPIRATION RATIO (g dry matter/kg of water)
Sorghum	180
Maize	146
Potatoes	98
Barley	95
Oats	86
Rapeseed	80
Rice	78
Peas	72

5.5.5 Too little water

If rainfall or the water retaining capacity of a soil is low, it may suffer from a water shortage in the summer. Lack of water will affect the yield and quality of most crops, but irrigation (watering) is expensive. A grower must balance the costs of irrigation which may have to be passed on to the consumer against the benefits of a higher yield, in order to decide whether to irrigate.

If the rate of evapotranspiration exceeds rainfall then the plant is said to be under 'water stress'. Water lost from the cells by osmosis causes the cells to lose their turgor pressure which results in wilting. The plant will try to conserve water by closing its stomata. As a result of this, the rate of photosynthesis is reduced and growth impaired. If these conditions continue for a length of time, the plant will die. Scientists have developed varieties of some species of plants which are able to survive drought conditions better than others.

5.5.6 Irrigation

Table 5.8 shows the average crop response to certain levels of irrigation. It costs the grower about £5 to put 1 mm of water on each hectare of land over the course of the year. This figure allows for labour costs, diesel costs, cost of irrigation licence and an estimate of the cost of the machinery required. All figures shown will vary depending on the year, the market and the amount of natural rainfall.

Table 5.8   The cost and effects of irrigation

CROP	IRRIGATION APPLIED (mm)	INCREASE IN YIELD (t/Ha)	INCREASE IN INCOME PER TONNE (£)	COST OF IRRIGATION (£)	PROFIT OR LOSS (£)
Cereals	50	0.5	150	250	−100
Cauliflower	50	7.5	3750	250	+3500
Lettuce	50	10.0	10 000	250	+9750
Peas	25	1.3	260	125	+135
Potatoes (early)	50	6.3	945	250	+695
Potatoes (main crop)	100	8.0	800	500	+300
Sugar Beet	100	5.6	180	500	−320
Raspberries	100	2.0	2000	500	1500
Grass (Hay)	125	2.6	130	625	−493

The figures give an indication of how the irrigation of each crop affects the yield. In all crops, irrigation will cause the yield to increase, but, in some crops, such as cereals, sugar beet and hay, the cost of irrigation to a suitable level is so high, and the increased income obtained is so small that it ceases to be worthwhile. However irrigation is useful in crops such as cauliflowers, lettuces and soft fruit.

Commercial irrigation of crops

In tropical regions, rain water which fell during the rainy season may be stored and used to irrigate some crops during the dry season. During the dry season there is often no rain for several months at a time, so irrigation may enable growers to increase yield significantly. Table 5.9 shows the effect of irrigating groundnuts in Nigeria.

Table 5.9   The effect of irrigation on groundnuts in Kano, Nigeria

YEAR	RAINFALL (mm)	RAINFED YIELD (kg/Ha)	EXTRA IRRIGATION (mm)	IRRIGATED YIELD (kg/Ha)	YIELD INCREASE (%)
1982	630	1614	160	3107	93
1983	432	580	110	3145	492
1984	507	950	115	1580	66
Mean		1048		2614	150

5.5.7 Too much water

In some areas of the world, crop growth is limited by too much water being available. The climate in New Zealand, where rainfall tends to concentrate between the months of June and September, means that waterlogging of soil is a major problem, especially on very heavy clay soils.

If soil becomes waterlogged, water fills the air spaces in the soil surrounding the roots so that root cells are unable to respire aerobically. Crops become stunted and discoloured and eventually, plants will die leaving water tolerant weeds to thrive. When plants die decomposition of dead root material will liberate toxic hydrogen sulphide. The rate of denitrification will also increase, removing nitrogen from the soil.

5.5.8 Drainage

On land where waterlogging is a major problem, growers may choose to install a drainage system. There are many different types available but they fall into two categories:

(i) **Surface Drainage** – surface drains or channels can be cut into fields or at the edge of fields to drain excess water. Surface drainage is suitable when the soil is of a type which would not provide a stable foundation for laying pipes, on moderately steep hills where soil erosion may present a hazard and in areas which would be difficult or expensive to clear e.g. woodland.

Surface drainage

(ii) **Subsurface Drainage** – subsurface drainage is more efficient than surface drainage and once laid leaves a field undisturbed. Subsurface drainage involves the laying of sloping perforated plastic pipes under the topsoil allowing water to drain away from the growing crop.

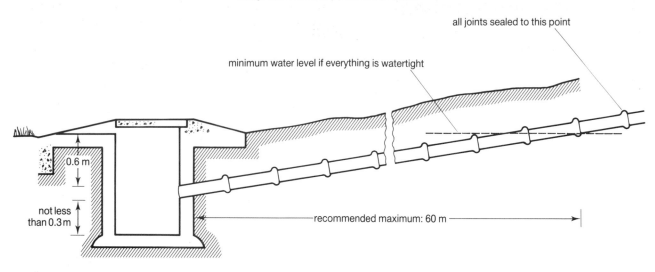

Longitudinal section of pit-overflow outlet

all joints sealed to this point

minimum water level if everything is watertight

0.6 m

not less than 0.3 m

recommended maximum: 60 m

Figure 41 Subsurface drainage

As with irrigation, a grower must balance the cost of drainage with the profits obtained from the increase in yields obtained.

5.6 NUTRIENTS

Apart from carbon, oxygen and hydrogen plants need at least 13 different chemical elements or nutrients to manufacture the whole range of complex organic compounds of which they are made. These elements are acquired by selective absorption from the soil solution (see Table 5.10).

Six of the nutrients are required in relatively large amounts and so are called **macronutrients**. These are the constituents of many plant components such as proteins, nucleic acids and chlorophyll. They are also essential for processes such as energy transfer, enzyme function and the maintenance of internal turgor pressure. The remaining seven nutrients are only needed in small or trace quantities and are referred to as **micronutrients** or trace elements. They have a variety of essential functions in plant metabolism.

For optimum growth, nutrients must be available in a soluble form, in appropriate and balanced amounts and at the right time in the growing season. Most agricultural soils contain considerable reserves of nutrients but these are largely tied up in the organic humus or as inorganic ions bound to colloidal clay and humus. Plants can only absorb nutrients as water soluble ions, so these reserves are mostly unavailable, although a small proportion are released into the soil solution each year through biological activity or chemical processes.

In natural ecosystems the processes of mineral recycling ensure that soil nutrient levels don't change very much from year to year, and although

Table 5.10 Plant nutrients

NUTRIENT	TYPE	REMOVAL WITH A WHEAT HARVEST (5t/Ha) (mg)
Nitrogen (N)	Macro.	105
Phosphorous (P)	Macro.	18
Potassium (K)	Macro.	15
Sulphur (S)	Micro.	8
Magnesium (Mg)	Micro.	6
Calcium (Ca)	Micro.	2
Chlorine (Cl)	Micro.	3
Iron (Fe)	Micro.	0.2
Manganese (Mn)	Micro.	0.2
Zinc (Zn)	Micro.	0.2
Copper (Cu)	Micro.	0.03
Boron (B)	Micro.	0.02
Molybdenum (Mo)	Micro.	Trace

the nutrient supply may limit productivity, the effect remains relatively constant. However, in agriculture and horticulture, the problems of nutrient supply are exacerbated by harvesting, which actively removes nutrients from the site. This results in the progressive depletion of soil nutrient reserves and the soil becomes impoverished. The soil nutrient supply becomes insufficient for crop needs and fertiliser application is the only way of maintaining productivity.

5.6.1 What are fertilisers?

Fertilisers help plants achieve their potential, freed from the limitation of nutrient supply from the soil. This is particularly important when high yields are required. Figure 42 shows the yield response of winter wheat to added nitrogen.

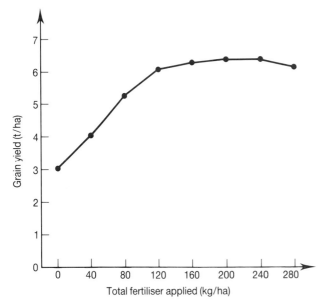

Figure 42 The response of winter wheat to nitrogen fertiliser

Fertiliser can be added in many different forms but are generally classified as organic or inorganic.

5.6.2 Organic fertilisers

Organic fertilisers consist largely of plant remains. The most widely used organic fertiliser on farms is farm yard manure (FYM). This is a mixture of cattle faeces, urine and straw, well mixed, and partly decomposed. Other forms of animal faeces such as horse manure and chicken droppings may also be used, and composts, leaf mould and peat substitute are widely used in horticulture. The use of organic fertilisers is a particular advantage on mixed farms.

'Green' manure is another organic fertiliser. This is a crop which is grown in the autumn on what would otherwise be bare land, in order to absorb the soil nutrients. The crop is ploughed back into the land and slowly releases the nutrients throughout the following spring. The crop is also known as a 'catch crop' and is a good way to reduce leaching.

Muck-spreading in action

5.6.3 Inorganic fertilisers

Inorganic fertilisers are mixtures of inorganic mineral nutrients. They may contain one or more of the macronutrients, in a form suitable for plant absorption, and sometimes micronutrients as well. It is normal to blend the mix of nutrients carefully to meet the specific crop requirement at the site. Table 5.11 shows the average nutrient content of organic and inorganic fertilisers.

Inorganic fertilisers

Table 5.11 The average N,P,K content of organic and inorganic fertilisers (percentage by weight)

	H$_2$O	N	P	K
Ammonium Nitrate	0.0	34	0.0	0.0
Triple Super Phosphate	0.0	0.0	19.5	0.0
Potassium Chloride	0.0	0.0	0.0	45
Farm Yard Manure	77	0.6	0.13	0.58

A farmer would need to combine the three inorganic fertilisers to provide the nutrients required. Inorganic fertilisers are, however, more concentrated and have much lower water contents, so the volume of inorganic fertiliser which needs to be applied is relatively low. A farmer applying organic fertiliser must spread large quantities of 'wet' manure to supply the required level of nutrition.

5.6.4 Plant nutrient supply

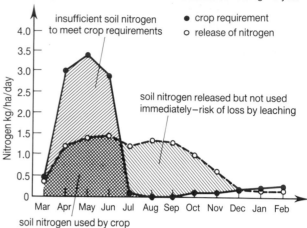

Figure 43 Crop requirement and nutrient release from the soil during the year

As you will realise the supply of nutrients to the growing plant is not constant throughout the year. The rate at which minerals are released from organic matter varies from season to season. Figure 43 shows that there are two peaks of release, caused by seasonal variations in microbial activity. The first peak occurs during April, May and June and supplies the growing crop with some of its nutrient requirements. However, to achieve optimum crop yields, additional nutrients are required, hence the need to apply fertiliser in the spring. The second peak occurs in September/October and is caused by soil cultivation which, together with warm temperatures, stimulates bacterial activity which results in the release of minerals into the soil solution. These minerals are subject to leaching.

Crops respond differently to different forms of fertiliser. Some plants are able to extract the nutrients they require more readily if they are supplied in a particular form.

5.6.5 Getting the level right

Pressures on farmers have never been greater. Falling agricultural prices and rising costs mean that *every* enterprise on the farm has to be operated efficiently to make economic sense. At the same time, there is increasing pressure on farmers to ensure that agriculture does not pose a threat to the environment. Fertilisers must, therefore, be used efficiently and with minimum impact on the environment. Before adding fertiliser to his land, a farmer must assess the nitrogen requirements of the crop, taking into consideration the existing nitrogen levels in the soil. This will depend on previous farming practices, e.g. the crops previously grown and the fertiliser previously used.

It is uneconomic to continue to add nitrogen to a growing crop, as its effects reach a plateau. This is shown in Figure 44. The yield of wheat increases with the addition of nitrogen, up to a maximum of 9 t/Ha. After this point, further addition of nitrogen had no effect on yield.

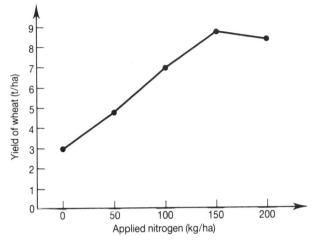

Figure 44 *The influence of fertiliser nitrogen on yield*

In cereals, the addition of nitrogen causes increased growth of the crop leaf as well as grain. When a cereal crop grows too rapidly it becomes susceptible to wind damage and lodging (laying down in the field). This will affect the harvestable grain yield as it is more difficult for the combine-harvester to pick up the ears and because the grain takes longer to dry out if it is close to the ground.

> **Now try Investigation 9 Investigating the Response to Different Levels of Fertiliser Application in B. Rapa in the *Plant Science in Action Investigation Pack*.**

5.6.6 The problems of nitrate residues in soil

If a large amount of nitrogen is added to a crop (as fertiliser), there is an increased risk of nitrate residues leaching into the soil. The relationship between the amount of fertiliser added and the amount of nitrate found in the soil is shown in Figure 45. All soils contain mineral nutrients, regardless of whether fertilisers have been used. They are formed naturally by the breakdown of organic matter in the soil (mineralisation) which occurs whenever the soil temperature is above about 5 °C. Any minerals left in the soil after harvest are at risk of leaching out of the rooting layer when it rains over the winter. The best way of keeping nitrogen leaching to a minimum, is to ensure that the least amount of nitrate possible is present in the soil over the winter. There are several ways of doing this:

(i) Ensure that the soil nitrate levels are as low as possible after harvest by making certain that the amount of nitrogen added matches the crop's demands closely.
(ii) Avoid applying nitrogen either as a fertiliser or as animal manure in the autumn and winter.
(iii) Keep a crop on the land over winter to mop up the remaining nitrogen.

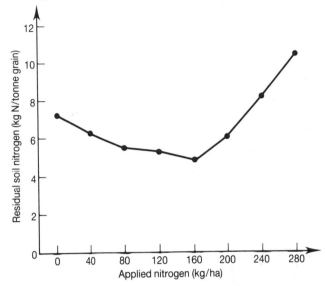

Figure 45 *The relationship between the nitrogen added as fertiliser and the residual soil nitrogen*

If a crop is to obtain maximum use of the nutrients supplied and minimum damage is to be caused to the environment, the timing of fertiliser application is vital.

5.6.7 The effects of nitrates leaching into the water supply

When nitrates are leached into the water supply it has two main effects:

(i) Risks to human health

In 1980 the EC agreed a limit of 50 mg/l of nitrate in drinking water. This standard is difficult to achieve in all places at all times, but particularly in lowland arable farming areas such as East Anglia. Here rainfall is low so there is less water to dilute the nitrate and concentrations tend to be higher than in the wetter west. Nitrate itself is not toxic, but when it is converted into nitrite by bacteria in the mouth, stomach and small intestine it can be.

These nitrites can react with the chemical **oxyhaemoglobin** which is found in the blood. Oxyhaemoglobin is formed when molecules of oxygen in the lungs, pass through the alveoli walls into the blood capillaries which surround them. The oxygen combines with haemoglobin, a red respiratory pigment, which enables the blood to absorb a larger volume of oxygen than would otherwise be possible. After oxygen is absorbed, oxyhaemoglobin becomes brighter red in colour. Oxygen is carried around the body in this form and released to the organs and tissues as it is required for use in cell respiration.

If nitrites are absorbed by oxyhaemoglobin, **methaemoglobin** is formed. This inhibits the carriage of oxygen by the pigment and so is potentially very dangerous as it reduces the amount of oxygen available within the body. This conversion is perfectly normal and there is a well established enzyme pathway in the body to reconvert methaemoglobin. Normally about 0.5–2.0% of the haemoglobin pool is found in the form of methaemoglobin. However, at increased nitrite levels this percentage may rise. Although this may cause no clinical distress, in newborn babies elevated methaemoglobin levels may lead to the so called Blue Baby Syndrome (infant methaemoglobinaemia). At levels of 10%, the skin will appear grey or blue in colour due to the lack of oxygen circulating within the body. Levels of 40% are considered to be dangerous. However patients have recovered from levels as high as 70%. Incidences of Blue Baby Syndrome are fairly rare – there have been no reported cases in the UK since 1972. It is therefore, difficult to draw direct links between this condition and elevated nitrate levels in the water supply.

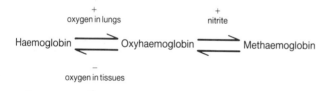

Figure 46   The effects of nitrite on haemoglobin

Nitrites have also been linked to stomach cancer. Nitrite is very reactive and may react in the stomach with food components to form carcinogenic compounds. However the evidence for this is inconclusive as the incidence of stomach cancer in the UK has been falling steadily over the last 20 years, whereas the level of nitrate in water has risen.

(ii) Eutrophication

Unicellular algae living in water like land plants, need the macronutrients nitrogen, potassium and phosphorous to grow. In certain conditions, high levels of nutrients will promote excessive (eutrophic) growth and give rise to so-called algal blooms. The result can be unsightly and bacterial decay of the algae may cause deoxygenation of the water at the end of the summer. In fresh water, it is mainly the presence of higher levels of phosphate from non-agricultural sources like sewage works which causes these problems.

Nitrates increase the growth of algae and plants in water

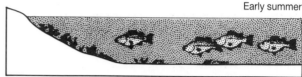

Plants grow rapidly and when they die are decayed by microorganisms. Increased plant matter leads to an increased number of microorganisms

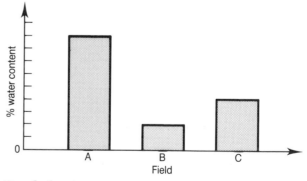

Increased size of microorganism population causes oxygen in water to be used up and cause fish to die

Figure 47   Eutrophication

> **Now try Investigation 10 Pollution of the Waterways in the *Plant Science in Action Investigation Pack.***

QUESTIONS

1 Imagine you own a farm with three distinct types of soil; sandy loam, loam and clay. The water content of three of your fields is measured during the winter. The percentage water is shown in the figure below:

Graph showing water content of each of three fields

a) Which type of soil is most likely to be found in which field? Explain your answer with reference to the soil's structure.

b) The soil in field A was found to be waterlogged. How do you think this affects crop growth?

c) The table below shows the field capacity and the permanent wilting percentage of the soil found in each field:

FIELD	FIELD CAPACITY %	PERMANENT WILTING PERCENTAGE %
A	40	20
B	10	3
C	28	12

   (i) What is meant by the 'field capacity'?
   (ii) What is meant by the 'permanent wilting percentage'?
   (iii) Use these figures to calculate the amount of water available to crops in each of the three fields.

d) The data below are temperature readings for two soil samples, one from field A and one from field B.

TIME (hours)	TEMPERATURE (°C) SAMPLE 1	SAMPLE 2
08.00	2	2
09.00	4	5
10.00	5	9
11.00	8	14
12.00	11	17
13.00	15	20
14.00	17	23

   (i) Plot the data onto one pair of axes.
   (ii) Work out which sample was taken from which field. Label your graph appropriately and explain your decision.
   (iii) Imagine similar readings were taken for field C. Sketch a third line onto your graph to show the likely position of the data obtained.

e) If you wished to improve the soil in field B so that it had a similar water holding capacity to the soil in field C what could you do?

2 The table shows the effects of increasing the level of irrigation on the yield of straw and grain in a crop of wheat.

The effects of irrigation on yield in wheat

Irrigation (cm)	12.5	19.0	38.0	63.0	125.0
Yield of grain (t/Ha)	2.5	2.7	2.9	2.9	3.0
Yield of straw (t/Ha)	3.1	3.5	4.1	4.6	6.0

a) Draw a graph to illustrate the figures in the table.
b) Comment on the response to irrigation of straw and grain.
c) When would it be worthwhile irrigating wheat?

3 The graph below shows the rates of transpiration and water absorption in an herbaceous plant over a 24 hour period:

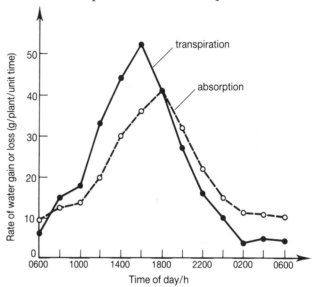

Graph showing water absorption and transpiration in an herbaceous plant over a 24 hour period

a) At what time of the day is the difference between the rates of the two processes at its greatest?
b) Account for the rate of transpiration
   (i) between 12.00 and 20.00 hours
   (ii) between 02.00 and 06.00 hours.
c) Suggest why gardeners are advised to transplant young seedlings in the evening.
d) The following methods are commonly used by students for measuring the rate of transpiration:
   (i) a potted plant with the soil covered with polythene is weighed at intervals;
   (ii) the rate of movement of an air bubble is measured in a potometer set up with a leafy shoot;
   (iii) single leaves are detached from a plant, suspended in the air and weighed at intervals.

Which of these methods gives the most accurate measure of transpiration rate? Give a reason for your answer.

(AEB 1991)

4 The graph below shows the amounts of maize grain produced when different amounts of a fertiliser were added to the soil.

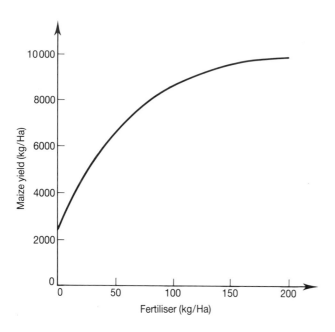

Graph showing maize yield

a) Approximately how much extra fertiliser was added to increase the yield from
(i) 3000–6000 kg/Ha?
(ii) 6000–9000 kg/Ha?
b) As a farmer, what would be the most economic quantity of fertiliser to add to the land?
c) What other environmental factors might affect the yield of maize obtained from a field?

5 a) Why is extra fertiliser required for crops in the spring?
b) Why is the nitrogen present in the soil in the autumn subject to leaching?
a) Prepare a leaflet for farmers warning of the dangers of nitrogen leaching and suggesting ways in which it might be reduced.

6 Most arable farmers in developed countries use inorganic fertilisers on their crops. The table below shows how the average yield of several crops has increased since 1963.

YEAR	WHEAT (t/Ha)	BARLEY (t/Ha)	OATS (t/Ha)
1963	4.16	3.61	3.89
1973	4.36	3.97	3.84
1983	6.31	4.63	4.38

a) What other factors have contributed to this increase in yields?
b) During the 1980s there was renewed interest in organic farming. What is meant by this term?
c) Evaluate the pros and cons of both organic and inorganic farming. Write a brief report recommending either one method or the other.

7 Gardening magazines urge us to treat our lawns with fertiliser. How would you test the hypotheses that:
a) the growth of grass increases with the amount of fertiliser applied?
b) the more frequently grass is cut, the more rapidly it grows.

8 Sugar beet (*Beta vulgris*) is a deep rooted crop grown on light soils in northern Europe for the high sugar content of its roots. Before the crop is sown, the land is ploughed to form a good seed bed. In addition, well rotted manure may be applied at this time. A few weeks before sowing, salt (NaCl) is applied; this provide extra chloride ions that the crop requires if it is to produce high yields of sugar. The seeds are small and are sown by direct drilling at the rate of 5–18 kg per hectare to give a final density of 80 000 plants per hectare.
a) State four factors which must be considered when deciding the density at which crop plants are to be sown.
b) Although applied to sugar beet in large quantities, chlorine is generally considered to be a micronutrient or trace element. Name one other trace element and state its role in the metabolism of a plant.
Fertiliser is applied when the sugar beet seed is drilled and again when the plants are at the seedling stage; pre-drilling applications of nitrogen fertiliser that were once quite common, are not now recommended. Rapid early growth is essential if high yields are to be produced; irrigation is often necessary during early growth, but not necessary later. The crop is harvested using mechanical harvesters. If the crop cannot be processed immediately, it is stored in pits covered with layers of straw and soil.

c) Suggest why it is recommended that nitrogen fertilisers should not be applied before drilling.
d) Explain why sugar beet may need irrigating in dry spells during early growth, but not later.
e) It is essential that 8% of the total yield of the crop is lost during harvesting. Suggest two reasons for this.
f) State two precautions that must be taken when storing a root crop such as sugar beet. Give a reason for each precaution.

(UCLES 1993)

BIBLIOGRAPHY

Chapman, S.R., Carter, L.P. (1976) *Crop Production.* W.H. Freeman.

Chrispeels, M.J., Sadava, D. (1977) *Plants, Food and People.* W.H. Freeman.

Forbes, J.C., Watson, R.D. (1992) *Plants in Agriculture.* CUP.

Hall, D.O., Rao, K.K. (1987) *Photosynthesis.* Edward Arnold.

Luckwell, L.C. (1981) *Plant Growth Regulators in Crop Production.* Edward Arnold.

McConnell, P. (1982) *The Agricultural Notebook.* Butterworths.

6 INTERACTIONS BETWEEN PLANTS AND OTHER ORGANISMS

LEARNING OUTCOMES

After studying this chapter you should be able to:

- assess the effects of pests and diseases on crop yields worldwide,
- define the term 'weed' and explain why weed species are so successful,
- describe the effects of interspecific competition between weeds and crops,
- outline the ways in which insect species can cause crop damage,
- list the many types of plant pathogens,
- outline the disease cycle and explain how the introduction of a small number of cells is sufficient to cause disease,
- classify the main pesticide types,
- outline the development of chemical pesticides, with special reference to the use of auxin based chemicals,
- assess the risks of using chemical pesticides to (i) the environment and (ii) human health,
- describe the use of 'natural enemies' and 'sterile partners' in biological pest control,
- list the physical methods which can be used to control pests,
- outline the technique of 'integrated pest management' and explain the implications of this method.

6.1 ASSESSING THE DAMAGE

Ever since man first started to cultivate plants and animals for food, there has been a conflict between man on the one hand, trying to establish a monoculture for efficient food production and nature on the other hand, trying to re-establish biodiversity. The cultivated field provides a large number of niches and there are always organisms, adapted to these niches waiting to fill them. Cultivated animals and plants are subjected to exactly the same attacks by other organisms when they are in cultivation, as they, or their ancestors, were in the wild. Similarly the cultivated field provides the ideal environment for the establishment of many of the coloniser plant species whose wind dispersed seeds are well adapted to reaching the fields. These uninvited guests of the agricultural system are responsible for enormous crop losses, by competing with the crop for resources, by consuming the crop both before and after harvest or by contaminating the crop and so rendering it useless.

Table 6.1   Estimated crop losses due to weeds, pests and diseases (%)

CONTINENT	PESTS	LOSS BY DISEASE	WEEDS	TOTAL LOSS
Europe	5	13	7	25.0
Oceania	7	13	8	27.9
North and Central America	9	11	8	28.7
USSR/China	11	9	10	29.7
South America	10	15	8	33.0
Africa	13	13	16	41.6
Asia	20	11	11	43.3

Table 6.1 shows the effect of weeds, pests and diseases on the yield of crop around the world. The risk of possible yield reductions on a large scale means the industry is prepared to invest a large proportion of its resources in crop protection. And this normally means the application of agrochemicals (Table 6.2).

Table 6.2   UK pesticide sales in 1990

	COST (£)	MASS (per 1000 kg)
Herbicide	181.2	11 927
Insecticide	38.6	1487
Fungicide	128.0	6756
Others	21.2	3580

The data in Table 6.1 shows that crop loss due to weeds, diseases and pests is higher in the Third World than it is in the Western World. If you take into account the thousands of pounds worth of pesticides that are used in the west to protect crops, this is not surprising.

Agrochemicals are extremely effective and they have an excellent safety record (British Agricultural Association 1991), but there is a growing public concern about the use of chemicals in food production and the cost is very high. As a result of these pressures there is increasing interest in alternative control measures, such as breeding disease, or pest, resistant varieties of a crop. The potential of gene transfer in plant and animal breeding is a rapidly expanding area of research. However there is also interest, particularly in the Third World, in improving cultural techniques, so that there is improved hygiene and seed cleaning, the use of crop rotations to prevent pest being carried over from one year to the next, and better timing of weed control operations. There is also a lot of interest in natural predators not only because it is often more satisfactory from an environmental point of view, but also because it often provides a more persistent level of control than a chemical spray. We will talk more about biological control in section 6.3.

6.2 WEEDS

A general definition of a weed is 'a plant growing in the wrong place'. Such a definition infers that any plant may become a weed, even crop plants can become weeds when a previous year's seed germinates in succeeding crops. However, a more important characteristic of weeds is that they interfere with the plants that are meant to be growing in a particular place. Interference may take many forms such as reduction in yield through competition, contamination of crops with poisonous seeds or berries, acting as a **green bridge** for pests and diseases, or even spoiling the visual appearance of an amenity area.

Whilst any plant may be considered a weed only a few plants are considered of major importance to commercial growers and farmers. According to Holm (1976) only 206 species can be regarded as important weeds and 80 of these are of major world significance (Table 6.3).

It is also interesting to note that three plant families; *Gramineae*, *Compositae* and *Cyperaceae*, contain almost half the world's major weeds.

Table 6.3   Some examples of flowering plant families contributing important weed species

FAMILY	NUMBER OF SPECIES	EXAMPLE
Graminoae	44	wild oats
Compositae	32	daisy
Cyperaceae	12	sedges
Polygonacaea	8	knotweed
Cruciferae	7	weed sugar beet

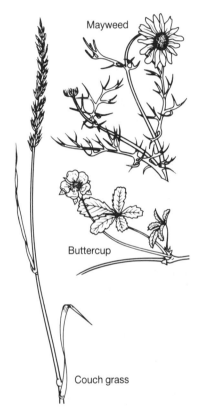

Figure 48   Common weed species – couch grass, buttercup, mayweed

Whilst weeds are generally regarded as harmful, they may also have beneficial effects. Their rapid colonisation of bare soil means they can quickly provide vegetation cover and so help prevent soil erosion. It has been recognised that weeds growing on field edges create a varied habitat providing an ideal environment for the maintenance of beneficial insects such as bumble bees, and predators and parasites of agricultural pests. Also, since most weeds are annuals and some are clearly related to crop plants (indeed many crop plants have originated from pre-historic weeds) they can provide a valuable source of genetic variability, including resistance to pests and diseases, which has proved useful in crop breeding programmes.

6.2.1 Why are weeds successful?

The introduction has suggested that certain plant families contain a large number of 'weed' species. What makes these plants particularly successful? There appears to be a number of characteristics which give plants the ability to thrive in a continually disturbed environment.

(i) Plasticity

Many weeds possess the ability to grow and reproduce under a wide range of environmental conditions. Where conditions are favourable, with adequate water and lack of competition they will produce large plants with a great number of seeds. If conditions are less favourable many weeds can complete their life cycles quickly producing smaller plants with fewer seeds.

(ii) Capacity for rapid increase

Weeds must possess the ability to colonise a new habitat rapidly. There are a number of weed characteristics which aid this process. Some weeds, when grown under favourable conditions, will produce vast quantities of seeds: Common Poppy (*Papaver rhoeas*), may produce 17 000 seeds, while a single Rosebay Willowherb (*Epilobium angustifolium*) can produce 80 000 seeds!

A poppy seed head can contain up to 17 000 seeds from a single flower.

Not all weeds are established from seed and there are many important weeds whose main method of reproduction is by vegetative means. A classical example of such a species is couch grass (*Agropyron repens*) where thin creeping rhizomes are easily fragmented and dispersed by cultural operations and this is its major method of spread.

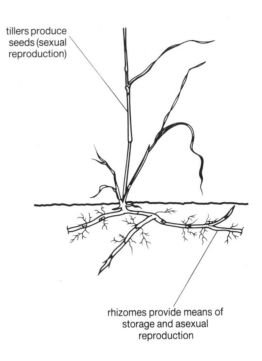

tillers produce seeds (sexual reproduction)

rhizomes provide means of storage and asexual reproduction

Figure 49   Reproduction, and spreading in couch grass

(iii) Discontinuous germination

By their very nature many weeds are well adapted to colonise disturbed land rapidly. One adaptation which enables them to do this is the production of seeds which will germinate only under certain circumstances. Many weed seeds possess dormancy mechanisms which allow them to survive in the soil until favourable conditions occur. A brief exposure to light may be a dormancy breaking stimulus for many weed seeds. This adaptation ensures germination when the soil is disturbed. Seed dormancy provides a powerful mechanism for weed survival and the duration of dormancy can last for many years providing a weed seed bank capable of exploiting any soil disturbance.

(iv) Intermittent germination

Intermittent germination from a persistent seed bank confers survival value in a disturbed habitat ensuring at least some plants flower and seed. The germination behaviour of weeds varies considerably with species (Figure 50) and knowledge of this can be useful in predicting possible weed problems.

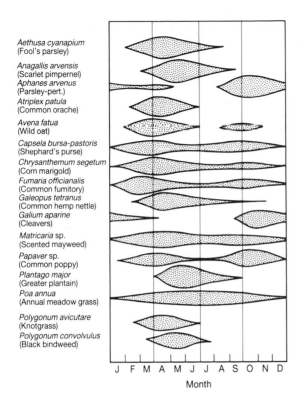

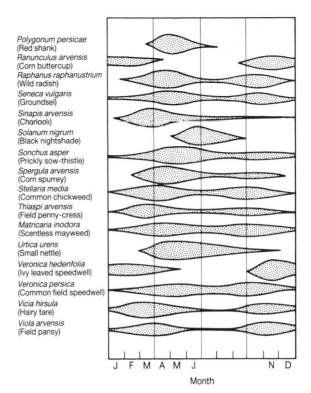

Figure 50 Main germination periods of some common annual weeds

(v) The ability to compete

Their capacity for rapid growth means that weeds can compete with crop plants for light, water and nutrients. Competition is interactive, with weeds suppressing the growth of crops, but crops themselves (many of which were originally selected from wild 'weed-type' species) will also suppress the growth of weeds. Because of this interaction, each crop has its own problem weeds. Clover (*Trifolium repens*) and chickweed (*Stellaria media*), which can be a problem in grassland are not generally a problem in the much taller growing cereals because they fail to compete successfully with the crop for light.

6.2.2 Weeds and competition with crops

The term competition may be used to describe one of three things:
  (i) the opposing demand of the various plant organs for assimilates;
 (ii) reactions to plant density which determine crop yield in pure stands (intraspecific);
(iii) the mutual interference between species which may occur in mixed crops (such as grassland) or between weeds and crops (interspecific).
Competition begins when the immediate supply of any necessary factor falls below the combined demand of the competitors. In other words, if two plant species occupy the same position and require the same resource, then they will compete with each other.

> **Now try Investigation 11 Which is More Competitive, the Crop or the Weed? in the *Plant Science in Action Investigation Pack*.**

6.2.3 Why are weeds such competitive plants?

A competitive plant is usually one which exhibits a high rate of growth during the seedling stage. Figure 51 shows the effects of weed growth on crop plants at different times of their life cycle.

a) Early season   b) Late season

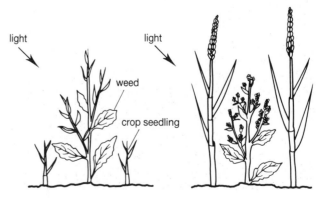

Figure 51 The effects of weed growth on a crop plant

If the weed is able to germinate and start to grow before the crop seedling is established then it is likely that the weed will overgrow the crop seedling and the crop will suffer from the effects of competition. If however, the crop plant is able to germinate and grow prior to the weed seedling becoming established, the reverse occurs, and the crop is more likely to survive.

> **Now try Investigation 12 How Does Sowing Density Affect Weed Development? and Investigation 13 What Type of Leaf Arrangement is Most Effective at Suppressing Weeds? in the *Plant Science in Action Investigation Pack*.**

6.2.4 What are the effects of this competition?

Weeds compete for water, light and nutrients but not necessarily all of these, all at the same time. Water and nutrients are drawn from an environmental reservoir, which holds an amount which may or may not be adequate for the needs of all the plants. Light, on the other hand, is immediately available and must be instantaneously intercepted or it will be lost as a source of energy for photosynthesis. The vital relationship in competition for light is one of physical position. The arrangement of the foliage for the interception of light energy, and especially its position relative to the foliage of competitors is critical. Simply, a competitive plant gets its leaves above those of its competitors.

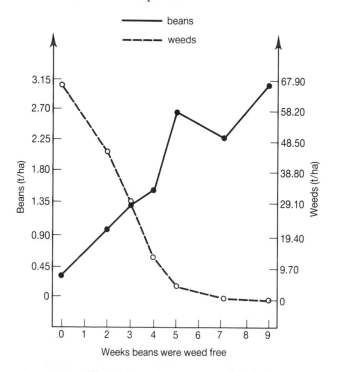

Figure 52   Effect of weeds on a crop of field beans

Figure 52 shows the effects of weeds on a crop of field beans. In general, as the density of weeds in the field increases, the yield of beans falls. The longer the field is left with weeds, the larger the proportion of the potential crop that is lost. As well as affecting the yield of the crop by competition, a weed population can cause problems at harvest, when weed seeds get mixed into the harvest crop. Contaminated samples of cereal crops must be cleaned before they can be used for milling, at extra cost to the grower.

It is often difficult to separate the effect of light from that of mineral nutrients since the nutrient availability will have a very pronounced effect on the plants' rate of growth. The species which can get nutrients more efficiently will presumably grow faster and get its leaves above those of the less successful species.

Water is usually only a problem – in temperate climates – in dry summers. In most years rainfall is adequate during the growing season. Table 6.4 shows the amount of water selected weeds and crops require for growth.

Table 6.4   The amount of water required to make 1.0 g of dry matter in three crops and three common weeds

PLANT		WATER REQUIRED (mls)
Crop	Oats	5.83
	Wheat	3.91
	Potatoes	1.37
Weed	*Spergula arvensis*	12.00
	Sinapis arvensis	5.00
	Polygonum persicaria	7.14

In general, weed species require more water to grow than crops. If the weed population is high, the water status of the soil will be fairly low and so less water will be available for crop growth.

6.3 INSECTS

There are an estimated 1.5 million insect species in the world and about 10 000 of these are classified as pests because they either damage livestock or crops, directly or indirectly.

(i) Direct damage occurs when adults or larvae feed on the crop.

(ii) Indirect damage occurs when adults lay eggs inside the plant (usually the flowers, fruit or woody stem) or by insects acting as vectors of viral diseases which damage crops.

6.3.1 Damage due to feeding

The damage done to crop plants by insects is mainly a result of their feeding habits. Insects can be divided into two distinct feeding types – the chewers and the suckers, as their mouthparts are adapted for particular functions.

Table 6.5   Some common insect pests

| ORDER | EXAMPLES | MOUTHPARTS | |
		LARVAE	ADULT
Orthoptera	Grasshoppers, Crickets	–	Chewing
Thysanoptera	Thrips	–	Rasping, sucking
Hemiptera	True Bugs	–	Piercing, sucking
Hymenoptera	Ants, Bees, Wasps	Chewing	Chewing, lapping
Coleoptera	Beetles	Chewing	Chewing
Lepidoptera	Butterflies, Moths	Chewing	Sucking
Diptera	Two winged flies, Mosquitos	Chewing	Piercing, sucking

Chewing insects, such as grasshoppers, eat their way through a plant riddling it with holes. Leaf chewers are sometimes unable to eat the tough vascular parts of the plant and so often leave a 'leaf skeleton' behind.

Sucking insects like aphids, have the ability to suck the sap from the inside of the plant due to their specialised mouthparts. This often leaves the growing plant stunted and deformed as nutrients are not able to reach the growing cells.

6.3.2 Damage due to egg laying

Although insects do not care for their young in the same way as some animals do, the female will lay her eggs in a protected site, such as inside the young wood of trees. This may cause the wood to split and eventually it will die. Emerging larva often cause further damage by feeding on leaves and bark. An example of this is the damage caused to apple trees by treehoppers.

Some insects lay their eggs inside the flower of a plant. This may affect fruit development as in the case of the strawberry weevil. This insect lays its eggs in the bud of a strawberry flower and then cuts a path through the stem so that the bud does not develop. This means that no fruit will develop.

6.3.3 Damage by virus vectors

Insects also act as vectors of some plant diseases, usually viruses. Aphids are particularly adept at transmitting viruses from infected plants as they move from plant to plant feeding on sap. The aphid feeds on the sap because it has a specially adapted mouthpiece which is able to pierce a hole in the plant tissue and so the sap is drawn up under pressure. If a plant is infected with a virus, the aphid will suck up the infected sap. When it moves to the next plant, there is a danger that the aphid will transmit the virus by contaminating the plant sap with the virus it is now carrying. The method of transmission is shown in Figure 53.

Leaf damage by grasshoppers can devastate a crop

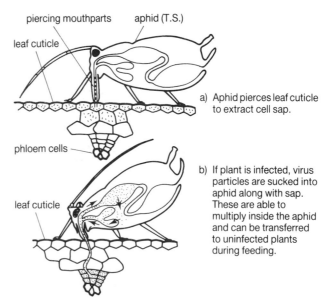

a) Aphid pierces leaf cuticle to extract cell sap.

b) If plant is infected, virus particles are sucked into aphid along with sap. These are able to multiply inside the aphid and can be transferred to uninfected plants during feeding.

Figure 53   Aphids feeding on sap can inject viruses into uninfected plants.

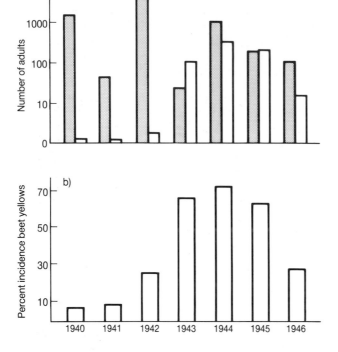

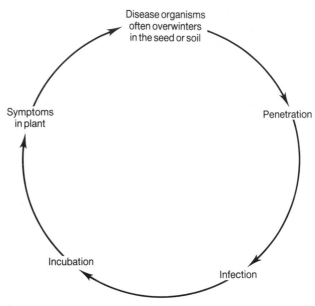

Figure 54   *The transmission of beet yellows by aphids*

The graph in Figure 54 shows the incidence of the beet yellows virus and the population size of two species of aphid. Beet yellows virus is a disease of sugar beet which can be transmitted by both the peach-potato aphid and the black-bean aphid. The black-bean aphid is found in large numbers but colonies tend to aggregate on only a few plants at a time. The peach-potato aphid is far more restless and although it is present in smaller numbers, colonies tend to move quickly throughout a crop. The peach-potato aphid appears to be most effective at transmitting the beet yellows virus as the incidence of the disease mirrors the increase in population size of this species. Although the peach-potato aphid is fewer in number its more mobile lifestyle means that it is able to transmit the virus more rapidly.

6.4   DISEASES AND THE DISEASE CYCLE

Crop losses due to plant disease exceed $3 billion annually in the USA. Disease is caused by pathogens which are either fungi, bacteria or viruses living as parasites on the crop. This is why pathogens often cause an infected crop to show signs similar to those of plants which are suffering from drought or mineral deficiency. All pathogens follow the same basic disease cycle as shown in Figure 55.

Figure 55   *The disease cycle*

The disease cycle begins with the **penetration** of the plant by the pathogen. Plants are usually well protected against pathogen penetration by the waxy cuticle that covers the stem and leaves. However, if the plant is damaged in any way the pathogen is better able to enter. Pathogens can also enter the plant through the stomata or with the help of insects, as discussed earlier in section 6.3.3.

Once inside the plant, the pathogen will cause an **infection** to occur. The parasitic nature of plant pathogens means that as they reproduce and grow, they compete with the host for water, minerals and the plant's metabolites causing the plant to produce **symptoms** such as wilting or yellowing. After the initial infection of the crop plant there is an **incubation** period, where the pathogen reproduces itself. As the pathogen numbers increase, the disease develops within the plant.

To prevent the rapid spread of a pathogen throughout the whole crop, a grower must treat the symptoms of the disease as soon as they begin to appear. This requires the careful monitoring of the crop for early signs of infection, taking into consideration the environmental conditions which can accelerate both infection and disease development.

6.4.1   Fungi

It is estimated that fungal diseases are responsible for yield losses in the region of 10–15% worldwide. They are difficult to control because the fungal life cycle is able to evolve rapidly, so overcoming the effects of many of the control measures used. Fungi can be divided into three groups as shown:

(i) **Phycomycetes:** causes downy mildew in grapes and blight in potatoes. (This group may be classified as three sub-groups in some texts.)

(ii) **Ascomycetes:** causes apple scab, powdery mildew and brown rot in peaches.

(iii) **Basidiomycetes:** causes rusts and smuts.

(i) Phycomycetes

Many common fungi belong to this group including the saprophytic bread mould *Mucor*. The most common agricultural pathogen belonging to this group is *Phytophthora infestans* which causes potato blight.

Potato blight

Potato blight is a fungal infestation which enters through the stomata and destroys the leaves, stems and tubers of potatoes. It is prevalent in Europe causing up to 16% of the total European crop to be lost each year. The fungi survives over winter in the tubers and when these are planted the fungus attacks the new shoots and leaves killing the cells. Potato blight favours high relative humidity (up to 100%) and temperatures of 16–22 °C, so warm, wet conditions are ideal. In order to prevent the spread of potato blight waste potatoes which have been discarded must be sprayed so that fungal spores are not produced. The potato crop can also be sprayed with a protective fungicide in the growing season.

Another common fungi belonging to this group is *Pythium debaryanum* which causes damping-off disease in cruciferous plants such as cabbage. *Pythium debaryanum* is a saprophytic/parasitic fungus which is particularly common on seedlings which are too crowded. It grows well in warm and moist conditions.

(ii) Ascomycetes

The most well known members of this group are the saprophytic mould *Penicillium* and the yeast *Saccharomyces*. One of the most common crop pathogens in this group is powdery mildew, which can affect several species. The photograph below shows the effect of powdery mildew (*Erysiphe graminis*) in barley.

Powdery mildew in barley

Erysiphe graminis is a parasitic fungi but its effects on the crop yield will depend on the time at which initial infection occurs. If the infection develops during the autumn, when the barley seedlings are very young, then the fungus causes the plant's roots to be stunted and so they are more likely to die over the winter. Those plants which survive winter are likely to produce fewer tillers and therefore, fewer ears. This will result in a marked decrease in yield. If the infection occurs later in the season when tillering is complete, the fungus will not affect the development of the plant itself. However the fungus will drain the plant of carbohydrate and thus affect the process of grain development. This again results in a reduction in yield as well as adversely affecting seed viability.

(iii) Basidiomycetes

Field mushrooms and puffballs belong to this group of fungi. So, however, do the fungi responsible for causing the diseases known as smuts and rusts in crops. Smuts and rusts cause the most damage to agricultural crops. They are able to enter the plant through the stomata and so are difficult to control.

The effect of rust (*Puccinia striifirmis*) in wheat is shown in the photo (page 64). Rust appears on the leaves and stems of the plant as orange streaks.

It is an airborne fungus, which favours cool, moist conditions (e.g. wet summers). Diseased plants may produce fewer, smaller seeds.

The effects of rust infection on wheat

6.4.2 Bacteria

Bacteria, due to their microscopic size, can easily enter plants through the smallest of wounds. Once inside a plant they multiply rapidly. One of the most serious bacterial pathogens is *Erwinia amylovora*. This is spread by insects as they collect nectar from apple and pear trees. The resulting disease, 'fireblight', causes blossom and shoots to die, therefore preventing fruit formation.

Fireblight

6.4.3 Viruses

Viruses are parasites which can only survive inside living cells. Once inside a cell they start to multiply, destroying it. They then gradually spread from cell to cell. Viruses are commonly transmitted by aphids but some are soil-borne. Infected plants often have yellow spots or streaks on their leaves, thus decreasing photosynthetic capacity and therefore, yield.

The beet yellows virus is a common cause of disease in sugar beet. It is transmitted by aphids and causes the leaves of the beet to turn yellow. This means that the beet is unable to photosynthesise at its normal capacity and so effects the yield of sugar which is built up in the root. Infected leaves are also more susceptible to fungal attack.

The effects of beet yellows virus

6.5 THE CHEMICAL CONTROL OF PESTS

Pesticide is the general name given to all chemicals designed to control pests. They are usually further classified according to the type of pest they control thus, they may be herbicides, insecticides, fungicides, rodenticide and so on. They are generally selective in their action but some pesticides control a wide range of pests.

6.5.1 The development of pesticides

Chemicals have been used for hundreds of years to help control pests. Marco Polo introduced the use of pyrethrum into Europe after observing its effects in the Far East. Pyrethrum is a natural insecticide produced by chrysanthemums. Before 1900 a small number of 'toxins' were used to control pests, such as copper sulphate, kerosene, mercury, arsenic and lead. However, these had obvious side effects and it wasn't until the 1940s that chemical pest control took off with the development of the herbicide 2,4-D (2,4-dinitrophenylhydrazine) and the insecticide DTT (dichlorodiphenyltrichloroethane).

Since the discovery of plant auxins in 1926, chemical companies had been investigating the use of auxins to improve yields in cereals. But by 1940 no useful effects had been observed. When the auxin NAA (a synthetic indole acetic acid) was added to cereal crops infested with weeds however, it was found that the weeds were killed and the cereals left untouched. NAA seemed to have a **selective** effect on broad leaved plants, however the dose rate was so high that it was too expensive to use commercially. By 1941, scientists had produced two synthetic auxins, MCPA (monochloropropyronic acid) and 2,4-D which selectively killed broad leaved plants at lower rates of application.

> **Now try Investigation 14 Using Synthetic Plant Growth Regulators as Herbicides in the *Plant Science in Action Investigation Pack*.**

Many of today's pesticides have selective action, allowing the grower to control specific pests whilst limiting environmental damage. Table 6.6 shows the relative toxicity of two insecticides.

Table 6.6   The relative toxicity of insecticide

| | PERCENTAGE KILLED | |
	HOUSEFLY (*Musca domestica*)	MUSTARD BEETLE (*P. cochleariae*)
Pyrethrin 1	1	80
Bioresmethrin	50	50

Pyrethrin 1 is based on the natural insecticide pyrethrum. Bioresmethrin is only slightly different in structure to pyrethrin. Pyrethrin 1 is a selective insecticide, having a great effect on the mustard beetle but little effect on the housefly. Bioresmethrin is non-selective and so has a similar effect on both insects. The non-selective insectide, like other non-selective pesticides will do less damage to the other species living within the environment, targeting only one or a small number of species.

6.5.2 Classification of pesticide types

There are many chemical pesticides on the market, however, they can all be classifed into one of the following groups:

(i) **Contact pesticides** coat the plant with a toxin during spraying. They are usually curative and are used to cure a particular infestation which is present in the crop. They are usually lower in cost than other types of pesticide but have a relatively short life span and so have to be reapplied. Contact pesticide enter the body of an insect through respiratory or sensory pores, whereas contact herbicides and fungicides are usually absorbed directly by the pest.

(ii) **Systemic pesticides** are absorbed through the roots and leaves of plants and transported around the plant. The plants sap becomes toxic to the pest and so when the crop is attacked by the pest the poison is taken directly into its system. Systemic pesticides tend to have longer term effects as they remain in the plant. They have a protectant effect on the crop as plants can be treated with pesticides to control pests which are likely to cause infestations before they occur.

(iii) **Residual pesticides** are added to bare soil before the crop is planted. These pesticides remain in the soil killing fungal spores, insect eggs or larva, and weed seedlings as they emerge. Residual pesticides have a protectant effect and must be fairly long lasting although ultimately any residue will be leached into the water supply.

6.5.3 Pesticides and food safety

The use of chemical pesticides is widespread in modern agriculture. Some of the benefits of using pesticides to control weeds, diseases and pests are:
- improved yields due to a decrease in crop failure or loss,
- improved quality of food due to the control of disease causing organisms,
- improved storage life of food,
- less time consuming than manual methods of pest control.

In the 1960s problems were discovered with DDT. It was found that it could remain in the food chain and actually cause genetic defects in some bird populations. Since these problems were recorded there has been great concern over the levels of pesticide which enter the harvested portion of the crop and the effects of these residues on human beings and other members of the food chain.

During the development of a new pesticide, many trials are carried out to assess its performance. These trials take the form of chemical tests, field tests, environmental tests, and toxicity tests. They can take five to ten years to complete before the pesticide is registered and released onto the market. During these tests scientists are trying to assess:

(i) What effect the pesticide has on the crop and the pest.

(ii) The effects of any other chemicals which may be produced when the pesticide is broken down in the soil and released into the environment.

(iii) The toxicity of the pesticide and any residue that might remain within the harvested crop.

Figure 56   The stages leading to the registration of a pesticide in the UK

6.5.4 Assessing the effects of pesticide residues in the environment

The breakdown of the component chemicals of a pesticide are monitored during trials by the use of radioactive isotopes. Radioactive isotopes exist for many biological atoms, and the most common isotope used is ^{14}C. This atom acts in exactly the same way as a normal carbon atom but as it is radioactive it is easy to detect. Radioactive isotopes can be introduced into the structure of a pesticide and the rate of breakdown can then be monitored by taking samples at regular intervals and testing for the presence of the isotope. The products resulting from pesticide breakdown can be determined by chemical analysis using techniques such as mass spectrometry and chromatography. Scientists will also study how the pesticide reacts with different soil types as this will affect its availability to plants and the rate at which any residues can filter through to the water table.

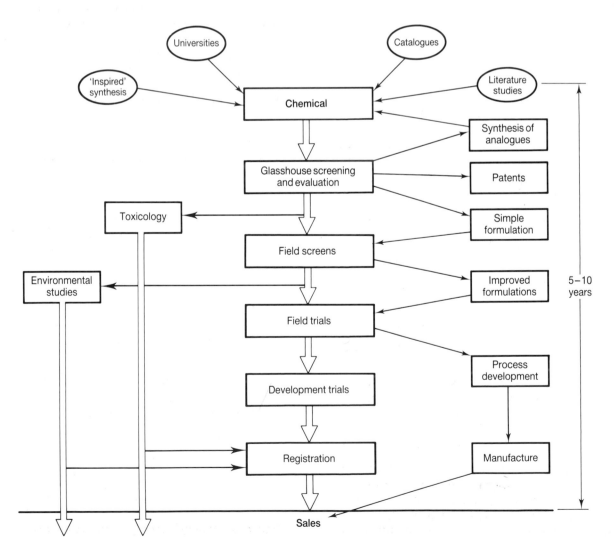

6.5.5 Assessing residues in harvested crops

It is important that scientists responsible for the development of pesticides understand how the chemicals they contain are metabolised by the crop and are able to identify any residues that remain within the harvestable part of the plant. Samples are harvested from test plants and prepared for analysis by grinding them finely. Solvents are used to extract the residues present. These are then measured using either gas – or high pressure liquid – chromatography. These techniques allow residue levels as low as 0.01 ppm to be detected. This is much lower than the level required to cause toxic effects and compares with finding a grain of salt in 100 kg of potatoes!

The Acceptable Daily Intake (ADI) level of pesticides for humans is calculated as 1% of the NOEL (no observed effect level). This is the level of a particular pesticide which causes no effect in toxicology trials. There is therefore, a hundred fold safety factor. Each food crop also has to contain less than the Maximum Residue Level (MRL). In many countries it is illegal to sell food with higher residues than this. In fact, the food that we eat contains a higher level of natural toxins than toxic pesticide residues, for example raspberries contain alkaloids which can cause liver damage if eaten in excess. Table 6.7 compares the risk of dying from pesticide poisoning with other causes of death.

Table 6.7   Risk of death due to pesticide poisoning

CAUSE OF DEATH	LEVEL OF RISK
Smoking cigarettes	1 in 200
Being overweight	1 in 600
Pesticide poisoning	1 in 750 000

6.6   THE BIOLOGICAL CONTROL OF PESTS

Good agricultural practices and the production of disease resistant strains are the most effective methods of biological control of pests.

6.6.1   Good practice

Pest and disease populations tend to build up over successive years and so **crop rotation** is a good way of ensuring that pest levels do not accumulate. The principle of growing a succession of different crops in the same field provides a changing competitive environment so that no one pest species can benefit from a constantly favourable environment. Rotation provides a means of living with pests which are never absent, but seldom serious.

Another method is to increase the density of sowing. An example of this is shown in Table 6.8.

Table 6.8   The effect of sowing density on the rate of complete canopy production in soyabeans

DISTANCE BETWEEN ROWS	DAYS TO PRODUCE COMPLETE CANOPY
100 cm	67
50 cm	47

If a crop covers the whole of the land it is growing on – if the crop canopy is complete – then weeds will be unable to grow as they will be unable to compete for light. If the sowing density of soyabeans is doubled, the time required to complete the crop canopy decreases and so the crop is less likely to be infested with weeds. Increasing the sowing density of plants does have several disadvantages: it is obviously more expensive and, in some cases some seedlings must be removed to allow the plants to grow and develop normally. This is done either by hand hoeing or using chemicals.

6.6.2   Natural enemies

In more recent years there has been great interest in using 'natural enemies' to control pests. This method is particularly successful in the control of insects and relies on the fact that most common pests are prey for other insects.

Aphids are common pests of gardens, greenhouses and crop plants. They are, however prey for hoverfly larva which possess piercing mouthparts and spear the aphid, sucking out its body fluids.

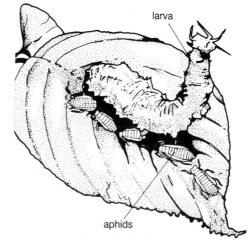

Figure 57   Hoverfly larva feeding on aphids

Adult hoverfly

In the 1980s, the Department of Entomology and Insect Pathology (at the Glasshouse Crops Research Institute) produced the following research data. Adult hoverflies were introduced into controlled environments containing aphid-infected cucumber plants. The adults were allowed to lay their eggs and were then removed. The larvae developed from the eggs and started to feed on the aphids. The results obtained are shown in Figure 58.

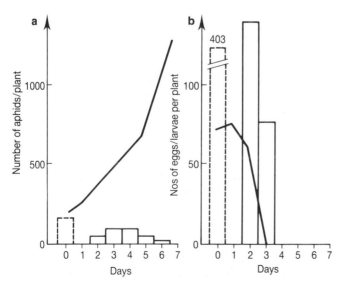

Figure 58   Graphs showing the effects of adding hoverfly eggs to a population of aphids

Graph **a** shows the effects of a small number of hoverfly larvae on the aphids. If the larvae population size is too small then the aphids are able to survive this method of control. However if the hoverfly larvae are introduced in sufficient numbers then they have a dramatic effect on the population of aphids.

This method of pest control would seem to have many advantages over the use of chemicals. However, it is an unsuitable method of control for many pests as it relies on the fact that a natural enemy exists for the pest. It is also almost impossible to use this method of control in the field environment.

> **Now try Investigation 15 Biological Control in the *Plant Science in Action Investigation Pack*.**

6.6.3 Sterile partners

Another method of biological control of pests involves the release of males which have been sterilised by irradiation treatment. This has been successfully used in the control of locusts and mosquitos. If sterile males are released into the environment, the females which mate with them lay eggs which are unfertilised and so do not develop. Eventually the insect population will be destroyed as adult insects are not replaced.

Adult locust

6.7 PHYSICAL METHODS OF PEST CONTROL

Many methods of physical pest control are expensive to employ as they may be selective or broad ranging. Some examples are discussed here.

(i) **Barriers** Screens or nets are often used in household gardens and horticulture to prevent destruction of crops by mice and birds. Young trees can be protected against damage to the bark by rabbits by guards which fit around the tree base.

Rabbit guards around young trees

(ii) **Heat** Seeds and bulbs can be protected against some pathogens by immersion in hot water (44–50°C) for short periods. The temperature of the water must be regulated so that pathogens are destroyed but the viability of the seed or bulb is not lost.

(iii) **Steam Sterilisation** is used in glasshouses to reduce soil infestations such as the spider mite.

(iv) **Bird Scarers** are often used in small fruit plantations.

Bird scarer

(v) **Traps** are used efficiently for rabbits in some areas. Light traps are also used in insect control.

(vi) **Weeding** Hoeing and other methods of cultivation are widely used throughout the world and are becoming more important in organic farming.

6.8 INTERGRATED PEST MANAGEMENT

Intergrated pest management (IPM) is a pest control technique which was developed in the 1960s. The main aim of this technique is to *control* the pest population at levels below those which cause uneconomic damage. Integrated pest management usually involves the use of non-chemical control methods such as natural enemies (biological control) or physical control, with the application of chemical pesticides only when the pest population exceeds a certain level.

When introducing this technique a grower must be aware of the types of pest which are likely to attack his crop and select a range of non-chemical methods to control them. This may involve the use of natural enemies. If these are to be encouraged it may be necessary to provide them with a suitable habitat such as replanting hedgerows, or an extra source of food. The pest population must be monitored carefully so that the population size does not reach levels which will cause crop loss. Although this is time consuming and can be difficult to assess, IPM results in a *decrease* in pesticide use which reduces costs and has environmental advantages.

QUESTIONS

1 a) Why is it an advantage for a plant to be able to complete its life cycle rapidly when growing conditions are unfavourable?
   b) Most weeds are self-fertile. How does this aid their survival?
   c) Why do weeds often produce such vast numbers of seeds?
   d) How is dormancy advantageous to a weed?
   e) How does weed dormancy cause inconvenience to man?
   f) How does intermittent germination affect the chances of survival in weeds?
   g) Why is intermittent germination not an advantage in crop plants?
   h) Why do clover and chickweed cause problems in grassland but not in cereal crops?

2 a) Why could the action of a leaf chewing insect affect the yield of a root crop such as beet?
   b) Why do sap sucking insects cause plants to become stunted?

3 In an experiment comparing the effectiveness of a particular insecticide against leaf-eating insects, samples of leaves were collected from treated and untreated parts of the same tree.

a) Give two precautions which should be taken in collecting the leaf samples in order to obtain statistically valid results.

The drawings below, which are life size, show three treated and three untreated leaves.

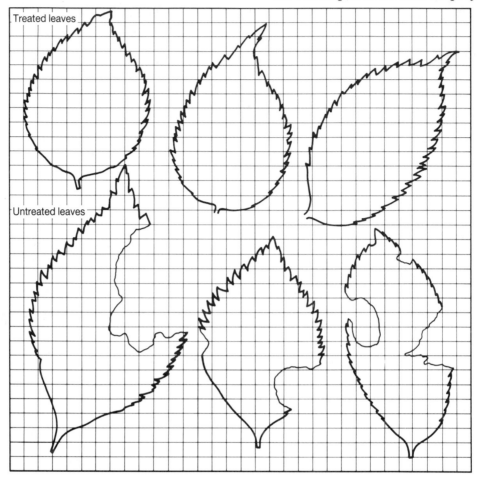

Leaves which have been treated with insecticide and leaves which have been left untreated

b) Make suitable measurements to compare the two leaf samples. Record the raw data in a table.

(AEB 1991)

4 Aphox and Metasystox are two commonly used insecticides which control aphids in cereal crops.

Table A	comparison of aphox and metasystox	
INSECTICIDE	COST PER HECTARE	SELECTIVITY
Aphox	£15.00	Aphids only
Metasystox	£7.50	Aphids, ladybirds, capsid wasps and others

What are the advantages and disadvantages of using each of these insecticide to control aphids in a cereal crop?

5 a) Why do systemic pesticides allow the protection of parts of the plant which have not developed at the time of spraying?

b) Why do contact pesticides have a relatively short life span?

c) Why are systemic pesticides more expensive than contact pesticides?

6 There is a lot of concern among the general public, over the use of chemicals to control pests in food crops. Imagine you are a scientist working for a large pesticide company. Prepare a short talk which will be delivered at a public meeting which:

- explains the advantages of using chemical pesticides on crop plants, with respect to the quality and yields of the product,
- explains the concerns which are commonly expressed about their use,
- assesses the risk of pesticide use to human health,
- draws a conclusion in which you state how far these concerns are justified and suggest what form of regulation, if any is required.

7 Blackspot is a fungus which parasitises roses. Infected plants are characterised by round, black spots which appear on the leaves. The drawings below show a number of infected leaves taken from plants showing different amounts of infection. In severe infections the leaflets may fall off and the plant may become completely defoliated.

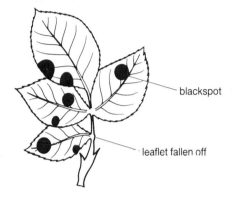

Leaves from rose bushes showing different degrees of infection with blackspot fungus

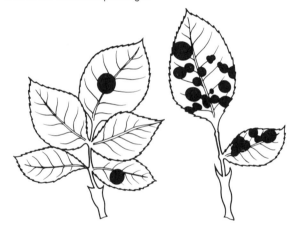

Leaves from rose bushes showing different degrees of infection with blackspot fungus

Since the growth of blackspot is inhibited by sulphur dioxide in the air, the amount of fungal infection may be used as a method of measuring the amount of sulphur dioxide in the air of a town or city.

a) Devise a scale which would give a quantitive estimation of the amount of blackspot on rose bushes at a specific place.

b) Describe a method you would use to sample the rose bushes for the distribution of blackspot in a small town.

c) (i) Suggest two factors, other than sulphur dioxide level, which might affect the incidence of blackspot at a particular site.
  (ii) How would you take these factors into consideration when planning your survey?

d) Explain why it would be important to apply an appropriate statistical test to the data you collect.

(AEB 1991)

8 Discuss the advantages and limitations of:
a) the use of natural enemies in crop production,
b) integrated pest management.

BIBLIOGRAPHY

Chapman, S.R., Carter, L.P. (1976) *Crop Production.* W.H. Freeman.

Dixon, A.F.G. *Biology of Aphids. (IOB)* Edward Arnold.

Fullick, A. *Death to Aphids. (SATIS 16–19)* ASE.

Hill, T.A. *The Biology of Weeds. (IOB)* Edward Arnold.

McConnell, P. (1986) *The Agricultural Notebook.* Butterworths.

Samways, M.J. (1981) *Biological Control of Weeds. (IOB)* Edward Arnold.

7 MODERN AGRICULTURAL PRACTICE – SOCIAL, ECONOMIC AND POLITICAL IMPACTS

LEARNING OUTCOMES
After studying this chapter you should be able to:
- compare the systems of cropping available to the grower and discuss the factors which must be considered when selecting an appropriate method of farming,
- discuss the impact of mechanisation and the use of agrochemicals on the farming industry,
- outline the aims of the Common Agricultural Policy (CAP), discuss its implementation and assess the success of this policy,
- discuss the effectiveness of using quota systems to control agricultural production,
- evaluate the effect of intensive agriculture on the natural environment.

7.1 RECENT CHANGES IN AGRICULTURE METHODS AND LAND USE

More than 80% of the land in Britain and Ireland is used for agriculture. Before the Second World War farms were small and usually family run. Traditional farms tended to grow a variety of crops in rotation and livestock was often kept to provide meat, milk and eggs, as well as a source of manure for fertiliser. The UK imported much of its food from abroad. However, during the war these imports stopped and farmers were encouraged to increase yields in an attempt to make the UK self-sufficient in several commodities. The change from traditional farming methods to more intensive farming methods and the increase in food production which has accompanied this has been made possible by:

- the development and increased use of inorganic fertilisers (see Chapter 5),
- the development and increased use of pesticides to reduce damage to crops (see Chapter 6),
- mechanisation – the development of machines which automated many of the formerly manual farming tasks (see section 7.3),
- the development of better varieties of crop which may be higher yielding, have better disease resistance or possess other improved characteristics (see Chapter 9).

7.1.1 Maximizing crop productivity

As we have seen from previous chapters crop productivity is ultimately determined by a combination of environmental factors including temperature, soil type, rainfall, nutrient availability and light. If a grower is to maximize the yield obtained from his crop, he must be able to assess how the natural environment meets the needs of his crop, and then determine the most economical ways to make up any shortfall. Modern developments provide a range of techniques, chemicals and machines which can be used to increase crop production, but the grower must balance the costs of such innovations both economically and environmentally.

Table 7.1 Increases in wheat yield in the UK since development of intensive farming methods

YEAR	WHEAT YIELD (10^5 t)
1965	3
1975	7
1985	12

Intensive farming methods have led to increases in yield (as shown in Table 7.1) and enabled farmers to produce cheaper food of a higher quality. However Government policy and new developments and techniques have encouraged farmers to over produce some commodities and legislation has since been introduced in an attempt to control production levels. Intensive farming has also had an impact on land use and the natural environment. The modern farming practices available, the legislation which controls production, and the impact of intensive agriculture on the environment, are discussed in this chapter.

7.2 CROPPING SYSTEMS

The cropping system used by a grower refers to the way in which he manages the growing of each crop from year to year. The system used by a grower will depend on factors such as soil type and climate and so may vary from region to region throughout the world. When selecting the most appropriate system a grower will aim to:
- maximize productivity,
- minimize variation within the crop from year to year,
- maintain soil quality and fertility.

7.2.1 Shifting cultivation

Shifting cultivation tends to be used in areas where soil has a low fertility such as the tropics, as it is the least intensive system of cultivation. Many tropical areas still practise the 'slash and burn' system of agriculture, where areas of forest are cleared to produce land for crop production. This land is initially very fertile due to the high organic matter content produced from forest leaf litter. However, after several years of continuous cultivation the land will become less fertile – fertilisers are often too expensive and therefore little if any is applied. Once trees are removed, land cleared in this way is often prone to erosion, especially if the region is subject to monsoons. After a few years the land becomes unsuitable for agriculture and is left fallow so that natural vegetation can return and replace the lost nutrients. This will take 10–20 years. Once one area has become infertile, the farmer will simply move on to another area and start the process of slash and burn again.

7.2.2 Continuous cropping

In many areas of the world shifting cultivation is impractical and uneconomic. Continuous or intensve croping techniques allow land to be cultivated year after year. There are several possible methods of continuous cultivation some of which depend on the use of agrochemicals and mechanisation and so may be inappropriate in some regions.

Areas of rainforest are cleared for agriculture by slashing and burning

Soil erosion is common on cleared land especially during the heavy monsoon rains

7.2.3 Mixed cropping or intercropping

Mixed cropping or intercropping is fairly common in the tropics. It involves the growing of two or more crops in the same field at the same time for example, maize and sorghum. The two crops may be sown simultaneously in distinct rows or with no distinct arrangement, or the second crop may be sown after the first crop has become established but before it is ready to harvest.

Intercropping has several advantages:
(i) The growing of two crops may insure against starvation if one crop fails. The risk of total crop failure due to attack from pests and diseases is reduced because it is

unlikely that an infestation of one type of pest will destroy both crops.

(ii) Intercropping allows a high plant population to exist. This ensures that more efficient use is made of resources such as sunlight.

(iii) Competition with weeds is reduced as the second crop acts as a desirable 'weed'.

(iv) Soil is protected from erosion because it is likely to be covered by vegetation continuously especially if the growing seasons of the two crops overlap.

(v) Standing crops can provide protection from the wind (wind break effect) and support, preventing the crop from being blown down (lodging).

(vi) If a legume (pea or bean) crop is grown as part of the system it will release nitrogen into the soil for use by the other crop.

The main disadvantage with intercropping is that it is difficult to manage processes such as harvest, especially if machinery is used. The crops are often mature at different times of the year or require different methods of harvest. This makes intercropping more suitable in regions where harvesting by hand is common.

Figure 59 *Intercropping of maize and cowpea in the tropics*

7.2.4 Monocropping

Monocropping or monoculture, is the use of a field to grow a sole crop. This type of cultivation is most common in developed regions of the world where crops are often grown as continuous monoculture year after year. If continuous monocropping is practised, growers need to use a high level of both fertiliser and pesticides in order to maintain high levels of productivity.

There are two reasons for this:

(i) Monocropping tends to encourage the build up of pests and diseases in the soil. If a different crop is grown each year pests specific to the first crop are unlikely to survive as the second crop often fails to provide a suitable host. Pesticides can be used to control this build up.

(ii) Monocropping of certain crops e.g. cereals tends to remove minerals from the soil which are not replaced unless nitrogenous fertilisers are used.

The main advantage of monocropping is that specialised techniques and machinery can be used for sowing, chemical application and harvest.

Monoculture of wheat

7.2.5 Crop rotation

Crop rotation allows a grower to keep his land in continuous cultivation without experiencing the disadvantages of growing a monocrop. This technique is widely used in both developed and developing regions of the world. An example of a four year crop rotation is illustrated in Figure 60.

Rotating crops means that pests which remain in the soil after the growth of one crop are less likely to survive the following year because their host crop is no longer present. If a legume (peas or beans) is included in a rotation the grower may not need to apply such high levels of fertiliser to the soil as the bacteria which live on the roots of legumes produce nitrates as a by-product of their metabolism. In some underdeveloped regions, the use of crop rotation may also lead to the production of a greater diversity of crops which lowers the risk of crop failure and increases variation in diet.

Year one

Field one (50 ha) Wheat	Field two (50 ha) Peas
Field four (50 ha) Sugar beet	Field three (50 ha) Wheat

Year two

Field one (50 ha) Sugar beat	Field two (50 ha) Wheat
Field four (50 ha) Wheat	Field three (50 ha) Peas

Year three

Field one (50 ha) Wheat	Field two (50 ha) Sugar beet
Field four (50 ha) Peas	Field three (50 ha) Wheat

Year four

Field one (50 ha) Peas	Field two (50 ha) Wheat
Field four (50 ha) Wheat	Field three (50 ha) Sugar beet

Crops are rotated on a four year cycle. Each year 50 ha of sugar beet, 50 ha of peas and 100 ha of wheat are grown. Rotation prevents the accumulation of pests in the soil e.g. sugar beet nematode. The inclusion of a legume e.g. pea replaces used nitrate.

Figure 60   Crop rotation

7.3   MECHANISATION

Since the industrial and agricultural revolutions the trend has been to increase the productivity of labour. In 1950 the work of one farm worker could provide enough food to feed 15 people. By 1975 this figure has risen to 55 people. Part of this increase in the productivity of labour is due to advances in plant breeding, pest control and fertiliser technology, but the increased use of machines has also had a significant effect.

Mechanisation increases crop production in many ways:
- deep ploughing allows better seed preparation;
- seed drills allow seeds to be planted at specified regular intervals resulting in a more uniform crop;
- harvesting is made easier by specific machines which rapidly pick and separate crops for storage;
- sprayer, spreader and irrigation systems allow growers to apply pesticides, fertilisers and extra water efficiently.

Some examples of agricultural machines

The increase in labour efficiency means that more land can be worked by fewer people (see Table 7.2).

Table 7.2   Man hours required to produce 100 bushels of corn (USA)

YEAR	MAN HOURS
1800	1000
1950	100
1960	10
1970	<3

7.4   AGROCHEMICALS

The demand for higher yields and improved quality has led to the development and use of several types of chemicals in agriculture. Agrochemicals commonly used in intensive agricultural systems include:

- fertilisers – which replace nutrients extracted from the soil during continuous cropping (see Chapter 5),
- pesticides – which control pests causing crop damage and disease (see Chapter 6),
- growth regulators.

7.4.1 Plant growth regulators

Throughout the history of agriculture, man has been modifying his crops in order to produce a better quantity and quality of food. Selective breeding has done much to 'improve' the genotype of domesticated animals and plants, but it tends to fix characteristics. A drought resistant variety may do very well in a dry year but be lower yielding than other varieties in a wet year. If the conditions are different from those expected, the farmer cannot change varieties mid-season!

However, a great deal can be done to affect the phenotype of the crop while it is growing. Plants in particular show very high levels of phenotypic plasticity and the grower can do much to alter the crops internal and external environment, so modifying the phenotype, to give improved yield.

In recent years there has been a growing interest in the possibility of modifying crop phenotypes of **plant growth regulators** (PGRs). These are organic chemicals which affect plant growth and development at very low concentrations. Plants manufacture their own plant growth regulators – the so-called plant 'hormones' or endogenous PGRs – and all metabolic activity within the plant is under their control. Nowadays most of the PGRs used in agriculture and horticulture are synthetic, although many of them are very similar to endogenous PGRs. The exception is the gibberellins, a chemically complex group, which are difficult to synthesise. Gibberellins, however, can be readily obtained from fungi (*Gibberella* sp) by fermentation.

Five individual groups of endogenous PGRs are known to occur in higher plants. These are the auxins, cytokinins, gibberellins, abscissic acid and ethene. With the exception of ethene, they are all found in the xylem and phloem saps, indicating vascular transport between organs. Whether this is evidence that direct messages are involved in coordinating the activities of the whole plant is not clear, but endogenous PGRs often have their effect at a site some distance from their source.

Synthetic PGRs are used in agriculture and horticulture for a variety of different purposes. They were first used commercially in the 1930s for promoting root initiation on cuttings. But since then, the use of PGRs has become increasingly sophisticated, and they are now used to control many of the developmental processes in plants. For convenience, current usage may be divided into four areas: vegetative propagation, control of dormancy, manipulation of developmental processes, and to facilitate plant breeding.

7.5   LEGISLATION AND THE CONTROL OF CROP PRODUCTION IN THE UK

Crop production is reliant on climate. This varies from year to year and so agricultural output is subject to years of surplus and deficit. In an open market the price obtained for a particular product will fluctuate considerably so that in years of deficit the price obtained may be high but in years of surplus prices will be low. In an attempt to prevent price fluctuations which affect both the grower and the consumer, the EC and the Government have produced legislation and a system of quotas and buying up schemes which control both the choice and quality of crops grown.

7.5.1 The European Community and the Common Agricultural Policy (CAP)

In 1957 six European countries signed the Treaty of Rome which set up the European Community (EC) and brought the Common Agricultural Policy (CAP) into action. In 1973 Britain and Ireland joined the EC and so were subject to the CAP legislation.

Table 7.3 Members of the European Community

YEAR	EUROPEAN COUNTRIES JOINING
1958	Belgium, France, Luxembourg, The Netherlands, Italy, West Germany
1973	Denmark, Eire, United Kingdom
1978	Greece
1986	Portugal, Spain
1995	Austria, Finland, Sweden, Norway

The Common Agricultural Policy (CAP) was designed to enhance the economic efficiency of the member countries and increase agricultural incomes by encouraging farmers to increase food production so that Europe could become self-sufficient. The main objectives of the CAP are:

- to increase agricultural production
- to assure availability of supply
- to ensure fair prices
- to stabilise agricultural markets
- to ensure a fair standard of living for agricultural workers.

The CAP attempts to meet these aims through a series of schemes:

- intervention
- common guaranteed prices for produce
- common preference for selling and buying produce
- grants for less favoured areas
- grants for environmentally sensitive areas.

7.5.2 The single internal market and intervention

If the CAP is to meet its objectives, free trade in agricultural produce between members of the community is required and imports from other countries must be controlled. This requires the removal of individual state support systems and their substitution with a common support policy which establishes a common price for produce. The principle commodities controlled by the CAP are:

- milk and milk products
- beef and veal
- sugar
- cereals.

These conditions account for 75% of all agricultural products. If these products are released onto the open market in years of surplus, the price obtained for these commodities would be low. The EC, therefore operated a buying up scheme where excess yield were brought and stored in order to maintain a stable price. This scheme where excess yield was brought and minimum price for these products. The products are stored and sold at a later date when supply is low and prices are higher. In theory this system should avoid excesses and shortages, and therefore stabilise prices, which is of benefit to both consumers and producers. In practice, the advantages in agriculture have led to overproduction and so the quantities of produce sold to intervention have been larger than expected. Overproduction means that there is never a shortage and so prices never rise above the intervention price. This means that the stores have continued to accumulate produce, resulting in the now infamous grain, beef and butter mountains.

7.5.3 Common guaranteed prices

Each year, the price of each commodity is fixed by the EC in Brussels. Each commodity is given a value in ECUs (European Currency Units). This is converted into each national currency according to the rate of exchange. This is so that the level of support received by a farmer from the EC is the same, irrespective of the country in which the farmer lives.

7.5.4 Common preferences

The prices for each product agreed by the EC are generally higher than the price of that commodity on the world market so that the farmer has a partly guaranteed income. If produce from the EC is exported to non member states, the EC pays an export subsidy to the farmer to make up the difference between the world price and the EC price. These subsidies are funded by money generated from the levies which are charged to members states which import non-EC produce, for example American bread wheat. This system of duties and levies helps to protect the European market from low price imports and ensures that priority is given to producers within the EC.

7.5.5 Other EC grants

(i) Less favourable areas (LFA)

Grants for farmers in 'less favourable areas' have been set up to improve the income of farmers who farm in environmentally poor areas. The grants paid are to compensate farmers for natural disadvatages and take the form of direct payments based on the size of the individual farm. Farmers eligible for a LFA grant will be farming land which has low fertility which cannot be improved chemically such as mountainous land, and will be producing yields well below the average.

(ii) Environmentally sensitive areas (ESA)

Payments of between £30 and £200 a hectare are available to farmers who farm their land in such a way that it conserves the landscape and the wildlife.

7.5.6 Controlling grain production – set aside

Overproduction of cereals in the early 1970s meant that there was a surplus of grain on the market. This caused the price to be deflated. To remedy this the intervention scheme was set up by the EC. As discussed earlier, the intervention scheme, whilst stabilising grain prices, resulted in the production of the now famous **grain mountains**, because so much grain was being sold into the stores. Overproduction of grain and a relatively high intervention price meant that grain stores kept expanding.

In 1993 new regulations were set up in an attempt to discourage overproduction of grain and other combinable crops, such as oil seed rape and peas. The so-called **set aside** legislation means that growers must take 15% of the land they use for growing combinable crops out of production. This land must not be used for animal or crop production but can be mown to prevent the production of weed seed heads which may affect the quality of subsequent crops.

In 1993 growers received a direct payment of 148 European Currency Units (ECUs) per hectare of set aside, and a further payment per hectare of combinable crops grown. These payments compensate for the expected decrease in the market price of grain due to changes in the intervention scheme. Figure 61 summarises intervention and set aside schemes. Table 7.4 compares the projected costs and income obtained per hectare of land growing wheat and under the set aside scheme.

Table 7.4   Set aside vs wheat production

		COSTS/INCOME (£/Ha)
Set Aside	Income	215.00
	Costs (mowing, ploughing)	30.00
	Profit	185.00
Wheat	Income	600.00
	Costs (chemicals, ploughing)	320.00
	Profit	180.00

These figures assume that wheat yields obtained are 6 t/Ha and it is sold at £100 a tonne. The data shows that this level of yield and price represents the break even point and if either

a) *Intervention*

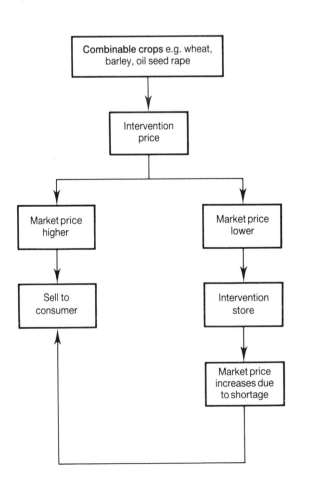

b) *Set aside*

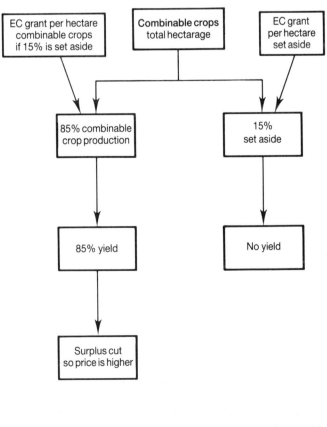

Figure 61   Intervention and set aside

expected yield or market price for wheat was lower than this, set aside would cease to be an economic proposition for the grower.

7.5.7 Agricultural policy in the UK

As members of the EC, farmers in the UK are subject to the common agricultural policy and its reforms. However, the supply of certain produce is further controlled in the UK by a system of **quotas**. A quota allows the farmer to grow a certain area or tonnage of a crop and if this level of production is exceeded the farmer may be penalised. Quotas are, therefore designed to discourage overproduction.

7.5.8 Growing potatoes – areas quotas

In the early 1970s, the Potato Marketing Board (PMB) was set up to control the supply of potatoes to the UK market. Initially potato growers were allocated a specific quota which allowed them to grow a certain number of hectares of potatoes. Each year growers pay a fee to the PMB for each hectare of quota they own (£100 a hectare in 1993). Growers can increase their quota by trading with other growers.

The PMB also controls the minimum size of potato which can be sold for human consumption. After harvest potatoes are sorted by passing them over a grid or riddle. The size of the gaps between the riddle rollers can be altered so that small potatoes fall through and only potatoes of a

certain size are selected and packed. In 1993 the minimum diameter was set at 45 mm. Potatoes under this size are dumped or used for animal feed. If there is an expected surplus due to good weather conditions during the growing season, the PMB will increase the minimum potato size so that a smaller proportion of the crop is selected.

Selection of potatoes for packing by use of a riddle

Some years yields are so high that despite the PMB controls there is a surplus of potatoes on the UK market. This has been made worse in recent years by the lack of control over imports from the EC. Since the development of the open European market, potatoes can be imported into the UK increasing supply and, therefore, decreasing the market price. In order to compensate for this, the PMB will sometimes operate a 'buying up programme'. In 1992 the PMB offered to buy surplus potatoes from growers for £40 a tonne. These potatoes are dyed with a non-toxic blue dye so that they cannot enter the human food chain but can be used for animal feed. The 'buying up programme' provides a minimum price for the market.

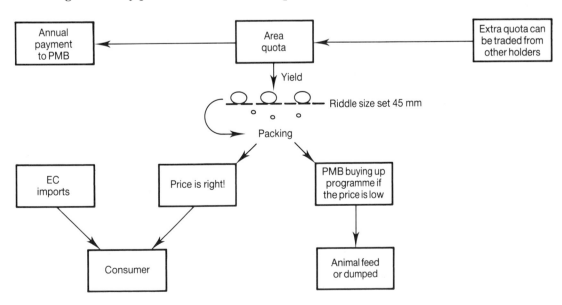

Figure 62   Control of potato supply by the Potato Marketing Board

7.5.9 Growing sugar beet – tonnage quotas

The supply of sugar to the UK market is controlled by British Sugar. All sugar beet grown in the UK is sold to British Sugar for processing. The supply of sugar from beet is controlled by means of tonnage quotas. Sugar beet tonnage is based on the sugar content of the beet being 16%. If the sugar content is higher than this, the tonnage will be altered accordingly. Each grower is allowed to grow a predetermined tonnage of sugar. This allocation is divided into three levels of quota:

- **A quota** – each grower is allocated a quota of a certain tonnage (e.g. 500 tonnes) on which they are guaranteed a predetermined price. In 1993 this price was £35 a tonne. A quota tonnage will be based on previous yields obtained by the grower.
- **B quota** – each grower is allocated a further 10% of their A quota tonnage (e.g. 50 tonnes) to fall into the B quota. The price of B quota sugar beet is usually the same as A quota but is subject to reductions if surpluses occur.
- **C quota** – all sugar beet grown over the A and B quota is put into the C quota category. The price of this sugar beet will be determined by the World Sugar Price and so is subject to fluctuation.

If a grower does not fulfil their A and B quota on a regular basis, they can be taken away.

7.6 THE IMPACT OF INTENSIVE FARMING ON LAND USE

Increased food production and the changes involved in intensive farming methods have had a great impact on land use in the UK. In order to produce more food two approaches have been adopted:

(i) Increasing productivity of existing farm land by the use of agrochemicals and improved crop varieties – these methods are discussed in other chapters.

(ii) Increasing the amount of land available for cultivation by clearing woodland, draining ponds and ditches and removing hedges.

Both of these approaches have had an impact on the countryside and the wildlife living there.

7.6.1 Woodlands

Britain was once covered by natural deciduous and coniferous forest. Much of this has been felled to provide wood for shelter, fuel and to provide raw material for paper making and to provide land for agriculture. By the mid 1800s only 4% of the UK was covered by forest. The removal of trees not only destroys the habitat of the wildlife which lives in the forest but will also ultimately affect the cycling of biological materials through the ecosystem.

Over the last 100 years, some woodland has been replanted and so woodland now covers about 9% of the UK. Trees replanted in this way tend to be planted and managed as a 'crop' and so tree plantations are often monocultures. Whilst plantations help to replenish stocks of raw materials, the original native woodland is of greater value to wildlife as it contains a wider variety of plants and trees providing food and habitats for a variety of animals. The main body to manage woodlands in Britain is The National Forestry Commission. They are now doing more to avoid monocultures especially of pine woodland by planting more hardwoods. Table 7.5 shows the rate of loss of ancient woodland in some counties of England and Wales.

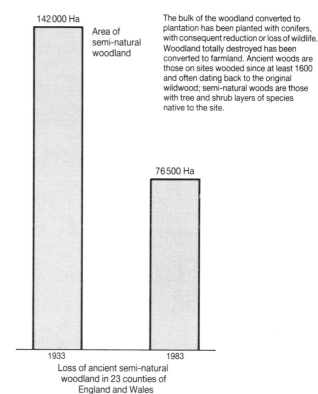

The bulk of the woodland converted to plantation has been planted with conifers, with consequent reduction or loss of wildlife. Woodland totally destroyed has been converted to farmland. Ancient woods are those on sites wooded since at least 1600 and often dating back to the original wildwood; semi-natural woods are those with tree and shrub layers of species native to the site.

Loss of ancient semi-natural woodland in 23 counties of England and Wales

Source: NCC Nature Conservation in Great Britain

Figure 63 *Map to show changes in areas of woodland in the UK 1933–1983*

The planting of trees on farms has been encouraged by the provisions of various grants. The Woodland Grant Scheme (WGS) introduced in April 1988, provides tax free planting grants and tax free proceeds from the sale of timber. The Farm Woodland Scheme (FWS) provides annual payments to farmers who convert productive farmland into woodland. Both schemes are

Table 7.5   Loss of ancient woodland in parts of the UK 1933–1983

COUNTY	ORIGINAL AREA OF WOODLAND 1933 (Ha)	SURVIVING AREA IN 1983 (Ha)	CONVERTED AREA TO PLANTATION SINCE 1933 (Ha)	AREA DESTROYED SINCE 1933 (Ha)	% AREA CONVERTED OR DESTROYED SINCE 1933
Avon	9367	2810	1040	141	30
Cambridgeshire	6167	2035	763	222	33
Leicestershire	3434	1614	1067	341	47
Northumberland	7176	3588	3194	323	50
Oxfordshire	15 711	5656	2884	270	36
Pembrokeshire	2392	1244	1286	44	52
Shropshire	6560	4133	6382	641	63
Surrey	11 492	4712	2428	840	41

(Nature Conservancy Council)

incentives to increase timber production, enhance the landscape and nurture the regeneration of lost woodland, thus providing habitats for wildlife. By promoting timber production, the WGS also provides an alternative to the normal range of crops grown and so may help to reduce surpluses.

7.6.2 Land drainage

Since Saxon times, man has increasingly managed wetland areas, ponds, ditches and rivers in the UK. Many farm ponds were originally dug to provide water for farm animals or habitats for domestic birds such as ducks. Ditches were dug as a means of draining wetland so that it was suitable for crop production. When drained, wetland areas, such as marshes, provide fertile land for crop production because they are rich in decomposing organic material. In some regions areas of bogland have been almost wiped out by the continual removal of peat for fuel. Waterways have also been polluted by industrial waste, sewage and agrochemicals. The problems associated with the pollution of waterways by fertiliser run-off are discussed in Chapter 5. The combined effect of these factors has resulted in the drastic reduction of wetland and natural water habitats leaving many species of animal and plant in danger.

It is estimated that half the field ponds which existed 50 years ago have disappeared. Table 7.6 shows the situation in Kimbolton, Bedfordshire. The data shows that the situation is getting worse with a 31% loss between 1890 and 1950 and a 52% loss between 1969 and 1980.

Extraction of peat for fuel has decimated some areas of bogland

Table 7.6   Changes in the numbers of ponds in Kimbolton, Bedfordshire

YEAR	NUMBER OF PONDS
1890	152
1950	103
1969	69
1980	49

(Nature Conservancy Council data)

The decline in areas of natural water and wetland has had a dramatic effect on the wildlife which lives there. This can be seen in a number of ways:

- Affects the population density of a species. For example, in the Huntingdon area of South Cambridgeshire the common frog population has declined from a level of 500 adults per 40 hectres in 1940 to just one adult per 40 hectares in 1974.
- Causes the extinction of some species. For example, 30% of all marsh plants identified in Bedfordshire in 1800 are now extinct.

7.6.3 Hedgerows

Many of the hedgerows found in the UK today are only one or two hundred years old, however some can be traced back to Saxon times. The Saxons originally built 'dead hedges' to provide boundaries between their land and that of their neighbours. These often needed replacing each year and so live hawthorn hedges were soon planted to replace them. Since Saxon times, many species have colonised ancient hedgerows and now most natural hedgerows consist of a range of both woody (elder, dogrose) and non-woody species (willowherb, ivy). The older the hedge, the more diverse it will be.

Hedgerows provide a physical boundary around fields, separating crops or livestock. They can act as a windbreak, protecting bare soil from erosion and standing crops from wind damage. Natural hedgerows provide a habitat and food for a wide variety of animals, including butterflies, and birds. They also can provide cover for game birds such as pheasant.

There are, however, several disadvantages associated with the use of hedgerows as boundaries.

(i) A hedgerows takes up land which could be used for crop production.

(ii) Plants growing near to the hedge will be shaded and so their growth will be stunted (see Figure 64).

(iii) Hedgerows also provide a habitat for some pests, such as weeds, insects and fungi, which could cause crop damage.

(iv) Maintaining a hedge is fairly time consuming and expensive.

(v) Large agricultural machines, such as combine-harvester, require a large turning area. The increased use of mechanisation in agriculture has created a need for larger fields.

A combination of these factors has led to the removal of many natural hedges. Table 7.7 shows the percent of hedges removed for agriculture between 1945 and 1972 in five English counties. This data shows that hedge removal is more common in areas where crop production is common than in areas where livestock or mixed farming occur.

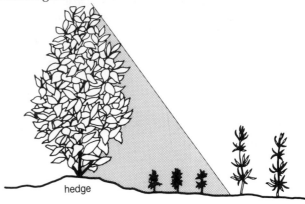

Figure 64 *Plants growing close to a hedge are shaded.*

Table 7.7 Hedge removal 1945–1972

COUNTY	HEDGES REMOVED %	MAIN AGRICULTURAL ACTIVITY
Huntingdonshire (now South Cambridgeshire)	38	arable
Dorset	10	livestock (dairy)
Herefordshire	10	mixed
Yorkshire	15	arable
Warwickshire	7	mixed

QUESTIONS

1 Ancient woodland provides a rich habitat for a large variety of species. The table below shows the area of ancient woodland lost in some counties in the UK between 1933 and 1983.

COUNTY	ANCIENT WOODLAND 1933 (Ha)	ANCIENT WOODLAND 1983 (Ha)
Cornwall	6348	3246
Essex	9596	7252
Humberside (formerly East Riding)	1194	767
Suffolk	4816	3022
Gwent	9691	3231
Hertfordshire	5984	3431

a) Calculate the percentage woodland lost between 1933 and 1983 in each county.
b) Suggest the reasons why so much woodland has been removed.
c) Some of the ancient woodland removed has been replanted with plantation trees. These are cultivated and eventually harvested for their wood. Why do you think plantation trees provide a poor habitat for wildlife?

2 Diagram A represents part of the water cycle as it relates to a large forest on level ground. The width of the arrows represents the approximate proportion of the water input going to different places.

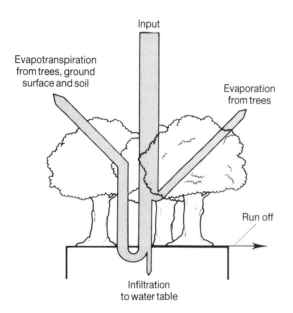

A *The water cycle within a forest*

a) Copy and complete diagram B to represent the fate of the same quantity of rainfall after deforestation.

B *The fate of water after deforestation*

b) (i) What major difference would you expect after deforestation, if the forest had been on steeply sloping ground?
(ii) Explain how removal of forest on steeply sloping ground could affect soil nitrogen.

(AEB 1991)

3 Tropical rain forest covers 10% of the Earth's surface. It is being destroyed at a rate of 6–10 million hectares a year mainly to provide land for agiculture or for wood production. Imagine you are an environmentalist. Write a short report to the Brazilian Government explaining the problems associated with slash and burn clearance of rain forest for agriculture.

4 The maps below show aerial views of Park Farm in 1950 and 1988.

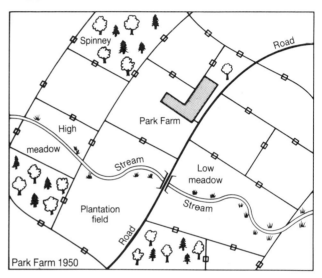

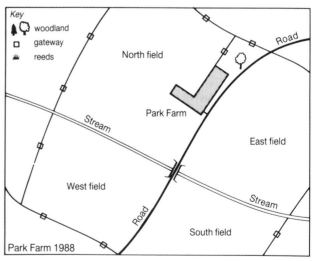

Park Farm 1950 and Park Farm 1988

State five changes which have occurred between 1950 and 1988. In each case explain the reason why this change is likely to have occurred and discuss its impact on the environment.

8.3 SOFT FRUIT

Worldwide each year 300 000 t of blackcurrants are grown. Of these 90% are grown in Europe with the UK being the largest producer worldwide. There are 4000 Ha of blackcurrants grown in the UK, mainly in the counties of Hereford and Worcester, Kent and Norfolk. Blackcurrants are mainly grown for juice extraction. During the Second World War imports of fruits from overseas were stopped and the population was encouraged to grow their own fruit. This led to a glut of soft fruit, including blackcurrants, which led to an increased interest in methods of processing soft fruit. It was discovered that blackcurrants contain large amount of vitamin C (between 196–242 mg per 100 ml juice, depending on the variety) and so processing plants were set up to extract the juice from the blackcurrant pulp. This is most often done by cold pressing. Pectinase enzymes are added to the blackcurrant pulp at 45 °C to break down pectin found within the primary cellulose cell walls. This causes the cell walls to collapse and allows juice to be released (see Figure 65).

PECTIN in the cell wall is made up of two chemicals

1. **Galacturonic acid** (GA) some of which is methylated ($^-CH_3$) 2. **Rhaminose** (Rh)

$$CH_3 \qquad CH_3$$

GA-GA-Rh-GA-GA-GA-Rh-GA-Rh

Pectinase enzymes will attack this structure

1. **Pectin methyl easterase** (PE) will attack the bond between the methyl group (C H_3 and Galacturonic acid (GA) 2. **Polygalacturonase** (PG) will attack the bonds between two galacturonic acid (GA) groups

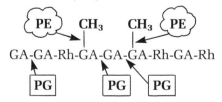

This results in the breakdown of the pectin molecule into small soluble molecules which can be washed out of the cell making it soften

Figure 65 The enzyme breakdown of pectin

Table 8.1 *The uses of blackcurrants in the UK*

USE	PERCENTAGE
Juice extraction	61
Jam production	15
Frozen	1
Canning and bottling	3

Blackcurrants are also used to produce flavourings such as yoghurts and the oil extracted from their buds can be used in perfume.

Table 8.2 *The Vitamin C (ascorbic acid) content of blackcurrants*

VARIETY OF BLACKCURRANT	VITAMIN C (mg/100 ml of juice)
Baldwin	242
Seabrook Black	205
Boskoop Giant	198
Weswick Choice	235
Mendip Cross	196
Wellington	204

> **Now try Investigation 16 Extracting Fruit Juice in the *Plant Science in Action Investigation Pack*.**

8.3.1 Growing blackcurrants

Blackcurrant bushes are perennial plants which survive for about 15 years. The plants require short days (less than 16 hours) before flowering is initiated. Flowers are produced in the spring, on a short stem which is produced as an offshoot of the main vegetative stem. After pollination the berries will develop from these flowers producing a crop of 5–6 t/Ha. The berries are usually harvested mechanically. After harvest the bushes are pruned. The wood which has produced fruit will be cut back so that new growth is encouraged. The old wood will produce fruit from season to season, however yield tends to decrease and mechanical harvesting is made more difficult if the bushes are allowed to grow too big. The main problem experienced by blackcurrant growers is the susceptibility of their crop to frost damage. This can be limited by spraying the flowers with water when they emerge which causes them to be more hardy and so less likely to be damaged by frost as described in Chapter 4. Problems with frost can also be avoided by selecting late flowering varieties so that flowering takes place when frosts are less likely to occur, or by selecting specially bred frost-hardy varieties.

8.4 FRUIT TREES

Worldwide each year 22.262×10^3 t of apples are produced. The UK produces mainly eating apples (18 000 Ha), some cooking apples (9000 Ha) and some cider apples (3500 Ha).

Apples are difficult to grow from seed because they have a juvenile period of 5 years during which time they will not bear fruit. Most apple trees are produced by techniques such as grafting. This is discussed later in this chapter. Trees are usually planted fairly close together so that they do not grow too large which helps when harvesting the fruit.

Apple trees are self infertile. They have to cross-pollinate as they possess an incompatibility system which means that pollen cannot be transferred from anthers to stigmas on the same plant or between plants of the same genotype. A mixture of trees must be grown together to ensure that pollination occurs.

Apples develop from the seed produced by the flower. The fruit must compete with the vegetative parts of the tree for carbohydrate to grow and trees can be encouraged to produce fruit early in the season if they are pruned to remove some of the vegetative parts. Fruit which cannot compete successfully for carbohydrate will drop off the tree during development. This is called 'self-thinning' or 'June drop'. This is illustrated in Figure 66.

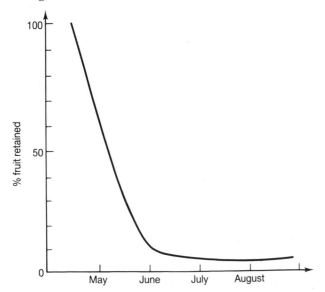

Figure 66 A graph to show the number of fruits in 'June Drop' or 'self-thinning' in apples

The dry weight of the apple is fixed early in the season. Any increase in weight later in the season is due to absorption of water. Apples respond well to irrigation. Self-thinning can be prevented by irrigation before June resulting in large numbers of small apples being produced per tree. However if larger apples are required, trees can be irrigated after the June drop.

As the fruit ripens the sugar content will increase and the skin colour will change. The colour of the skin is due to a pigment called **anthrocyanin** and its production is stimulated by light. Apples are still 'alive' when they are harvested. During storage their dry weight will drop due to continuing respiration. To prolong storage and shelf life, the respiration rate must be reduced. This can be achieved by reducing the oxygen level within the store or reducing the temperature.

8.5 GLASSHOUSE FRUIT

The tomato crop is the most important crop worldwide. Approximately 54×10^6 t are produced annually worldwide. In the UK there is a demand for both fresh and processed tomatoes. Tomatoes for processing (tinned tomatoes, tomato sauce) are imported, but fresh tomatoes are produced under glass between February and October. Outside this period they are largely imported.

8.5.1 The commercial production of tomatoes

A typical production cycle used by a commercial tomato grower might be as follows:
  (i) Seeds are planted in seed tray in November.
  (ii) The seedlings are transplanted into a growing medium and allowed to grow and mature.
  (iii) Fruit is produced by either flower pollination or parthenocarpy.
  (iv) As the fruit matures it will change colour from green to orange to red.
  (v) The tomatoes are harvested when they are orange as it takes 5–7 days for them to reach market. By this time they are fully ripe.

Fruit is produced on an **inflorescence** which is produced by the main stem. Flowers are self pollinated, however this can be encouraged by shaking. Fruit produced as a result of self-pollination are seeded. Non seeded fruit can be produced by inducing **parthenocarpy**. This involves the use of the hormone auxin to induce fruit development without pollination (see Chapter 1). This process allows fruit production to occur throughout the year.

The tomato flower is self pollinating

Like apples the dry weight of the tomato is determined early on in the growing season. Further growth is due to the uptake of water and the expansion of the cells already present. It is important that water is supplied to the crop regularly as if it is suddenly stopped the outer wall of the tomato looses its plasticity and is more likely to crack. As these cracks heal, they form cork cells which decreases the quality of the tomato.

The crop yield will depend on the time of harvest. Early in the year plants produce low yields of tomatoes but they fetch a higher price.

8.5.2 Ripening tomatoes

As with many horticultural products, it is essential that tomatoes ripen at the correct time in order to ensure that they can be sold at a good price. Tomatoes are **climacteric** fruit. This means that ripening is accompanied by an increase in respiration rate which uses up dry matter and so reduces the weight of the crop. In order to limit this loss, ripening is controlled in the tomato store after harvest. This can be done in one of three ways:

- Temperature control. Reducing the temperature of the store will slow down ripening. However temperatures below 10°C will cause chill damage and so this method is limited to the removal of **field heat**. The internal temperature of the tomato is higher than the temperature of the surrounding air. This internal temperature or field heat decreases as temperatures drop towards evening. If tomatoes are picked during the evening they will have a lower temperature. Field heat can also be reduced by cooling the fruit in water after picking.

- Light intensity. The rate of ripening is reduced if tomatoes are stored in the dark.
- Hormones. Ethylene is produced by the fruit prior to ripening. If this can be inhibited or supplied, ripening can be controlled.

8.6 GLASSHOUSE VEGETABLES

Lettuces are the second most important greenhouse crop in the UK. They are very delicate plants which bolt (flower) and rot easily. There is now a demand for lettuce all year round, and so seedlings are often propagated in greenhouses and transplanted into the field as small plants. Transplantation allows the grower to select the most uniform plants and the most suitable sowing rate. It also allows lettuce production to begin earlier in the year, producing an earlier crop which will sell at a higher price. The main disadvantage of transplanting seedlings is that it is very labour intensive and that the roots of the plants are easily damaged. Some lettuces are grown completely under glass and this obviously raises their cost.

Lettuce seedlings are often transplanted into fields.

8.7 FLORICULTURE

Floriculture – the production of flowers under glass – is a highly competitive and technological business. The production of flowers such as chrysanthemums and carnations, all year round requires precise environmental control. The chrysanthemum is the largest selling flower crop worldwide. The main reason for this is that it is easy to produce flowers all year round by manipulating the amount of the light the plants receive.

Chrysanthemums are short day plants. This means that they will normally flower during the autumn and winter. This response to day-length is controlled by the phytochrome system which is discussed in Chapters 1 and 3. During the winter plants are exposed to short periods of light and long periods of darkness. This causes the pigment P_R to accumulate and flowering to be initiated. If a grower wishes to prevent flowering during the winter he must provide supplementary lighting. It has been found that if the long winter night is interrupted by a short period of light, it has the effect of breaking the night into two short periods of darkness. This means that the chrysanthemum does not receive the long period of darkness required to cause P_R to accumulate and so will not flower.

Now try Investigation 17 Increasing the Flower Life of Daffodils in the *Plant Science in Action Investigation Pack*.

8.8 PROTECTIVE CROPPING

The use of a greenhouse avoids adverse environmental conditions and so allows the production of crops out of season for commercial and experimental use. Greenhouse production is

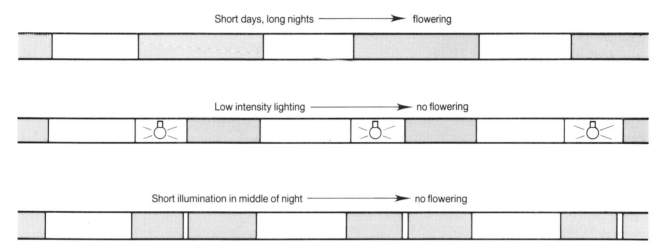

Figure 67 The use of supplementary light to prevent flowering in chrysanthemums

If a grower wishes to produce chrysanthemums for sale throughout the year, then he must manipulate the day-length and period of darkness that the plants are exposed to during the summer. This can be done by covering plants and thereby lengthening the period of darkness they receive.

fairly expensive and so it is only suitable for the production of the more expensive crops.

8.8.1 Controlling the environment

The greenhouse industry probably started during the 1600s in the Netherlands when Amsterdam was the world's biggest commercial city. Nowadays glasshouses are used extensively in the production of cut flowers, flowering and foliage plants, and fruit and vegetables. One of the main features of glasshouse production compared with the field, is the control of growth by manipulating the climate. Energy (as solar radiation) and carbon dioxide can be supplied and by the use of hydroponic cultures, or substrates like peat or rockwool, control of the root environment is possible. The use of computers for the accurate control of these factors has been widely adopted by growers. Since running costs of a glasshouse are high, a quantitative approach to management is rewarding.

Chrysanthemum production

Glasshouse crops

Table 8.3 Energy capture and plant production in symmetrical and asymmetrical greenhouses

| | AVERAGE TEMPERATURE °C | | TOTAL PLANT MASS (g) | |
	ASYMMETRIC	SYMMETRIC	ASYMMETRIC	SYMMETRIC
6 December – 15 January	15.5	14.5	2469	2217
17 January – 30 January	16.5	15.5	3556	2728
2 February – 24 February	17.0	15.5	10 009	9205

8.9 LIGHTING THE GREENHOUSE

8.9.1 Natural light

In most greenhouses, natural light inside the greenhouse is less than half that outside. The amount of light trapped by the greenhouse can be increased by changing the shape of the roof from the traditional symmetrical shape as shown in Figure 68, to an asymmetrical shape, as shown in Figure 69. The asymmetrical greenhouse roof is made up of several sheets of glass held at different angles. This increases the amount of sunlight trapped so increasing plant growth as shown in Table 8.3.

8.9.2 Supplementary light

Poor natural light in winter may be supplemented by artificial light from electric lamps. Ordinary household bulbs do not provide sufficient light intensity for photosynthesis, so high-pressure mercury vapour lamps are widely used in glasshouses as they produce a bright light with a spectrum close to the action spectrum for photosynthesis. Sodium vapour lamps are used increasingly as supplementary lighting as they are more efficient at converting electricity into light. Unfortunately the spectrum they produce does not match closely with the action spectrum for photosynthesis.

Supplementary lighting to aid photosynthesis is considered economic only when plants are small and therefore close together on the bench, making it possible for each lamp to illuminate many plants.

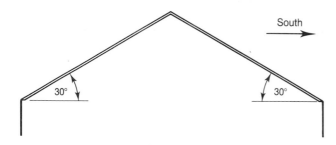

Figure 68   Traditional symmetrical roof

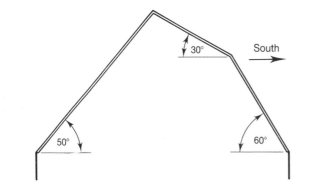

Figure 69   Asymmetrical greenhouse roof shape

Supplementary lighting in the greenhouse

Plants can be grown perfectly well without any natural daylight, as long as they receive sufficient intensity of artificial light. This can be provided by closely spaced florescent tubes in a special *growth room* or growth cabinet, in which other essentials such as heat and water are also provided.

> **Now try Investigation 18 The Rate of Photosynthesis in the *Plant Science in Action Investigation Pack*.**

8.10 TEMPERATURE CONTROL

The lowest (minimum) temperature can be controlled by a heating system; the highest (maximum) temperature can be controlled by ventilation and shading.

Maintaining a minimum temperature is relatively straight forward. Heaters may either be electric or, more commonly in large commercial glasshouses, oil or gas fired boilers heating a hot water system with radiators, not dissimilar to a domestic heating system. The minimum temperature is set on a thermostat and the heating switches on whenever the temperature falls below the minimum.

It is rather more complicated to maintain a temperature below a maximum as venting affects both temperature and humidity. Venting is necessary in summer when temperatures in the glasshouse can rise dramatically: the greenhouse effect. In most situations opening roof vents allows the hot air to escape and be replaced by cooler air through vents near the ground. Provided the air keeps moving through the glasshouse, temperatures can be kept within reasonable levels. In some situations it may be necessary to employ fans to assist with moving the air.

Ventilation systems in a glasshouse

However, when bright sunlight is to be expected during the summer, glasshouses are often also equipped with shading screens. These are sheets of reflective material which will cut down the amount of light, and therefore heat, reaching the crop. They can be drawn across the roof of the glasshouse whenever the light intensity exceeds a specified level.

One drawback to venting is that it will tend to reduce humidity. Warm dry air increases the rate of transpiration and may encourage certain pests, such as red spider mite which commonly attacks crops such as cucumber and tomatoes. All the conditions which increase transpiration tend to occur in glasshouses at the same time. Thus during warm sunny weather when the vents are open it is necessary to try and increase humidity. This can be achieved by damping down, misting, and partially closing the vents. However the rise in temperature that will occur in the glasshouse due to the closure of vents, will tend to reduce the relative humidity.

8.11 CARBON DIOXIDE

Carbon dioxide is an essential raw material required by plants for photosynthesis but it is only present in the air at about 0.03%. In a closed glasshouse this small amount can be even further reduced by the active photosynthesis of large plants. Ventilation is a partial solution but experiments have shown that by increasing the amount of CO_2 in the air (i.e. enriching the atmosphere) plant growth is speeded up and quality of the final crop improved. However CO_2 enrichment will not compensate for poor growing conditions such as low temperature, poor light or lack of water and nutrients.

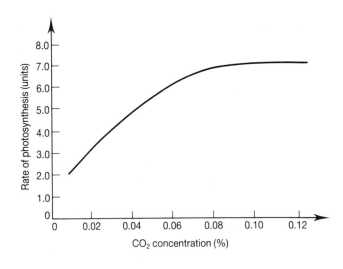

Figure 70   A graph to show the rate of photosynthesis and carbon dioxide concentration

CO_2 concentrations are normally increased to about two or three times the normal levels (0.06–0.09%). There is evidence to suggest that plants can take advantage of CO_2 levels up to 1.0%, provided that other factors are not limiting but that above this level, there is some toxic effect.

For obvious reasons CO_2 enrichment is usually only carried out in the light and when the vents are closed. Open ventilators would allow the CO_2 enriched air to escape. Enrichment is therefore only worth while for most crops during winter or early spring when the ventilators do not have to be open much to control the temperature. However with additional CO_2 the temperature can be allowed to rise higher before ventilation is given, as plants can make better use of the extra CO_2 at temperatures of 18–21 °C. With tomato and cucumber crops, the uptake of CO_2 in the summer is so rapid that summer enrichment with carbon dioxide up to levels of 0.06% have been found to be economically viable.

Enrichment is achieved either by releasing pure CO_2 into the glasshouse from dry ice or from a gas cylinder or it can be formed by burning a fuel such as paraffin and allowing the waste gas into the greenhouse. In theory the exhaust gasses from the heating boiler, which would normally be pumped into the atmosphere, could be used but this is seldom a very pure source and many toxic and oily products would also be present.

Carbon dioxide enrichment

8.12 CONTROL OF THE ROOT ENVIRONMENT

Apart from anchoring the plant in the ground, the main function of the root system is to absorb water and nutrients from the soil. Roots can only carry out this function if air is also present to supply

oxygen for cell respiration. The amount of water and nutrients added to the soil can be controlled, but the amount of air depends on soil structure (see Chapter 4).

Problems in growing crops in the soil arise from poor soil structure, variable water table, build-up of soluble salts, and pest and disease problems arising from repeated monocropping. In the UK and elsewhere, cropping in isolated containers of peat compost or a peat substitute has been adopted, but, because the volume of peat per plant is kept low for economic reasons, problems often arise in ensuring adequate and uniform provision of water and nutrients to the crop.

8.12.1 Nutrient film techniques

Recently growers have developed a process known as nutrient film technique (NFT), which offers the possibility of very precise control of the root environment. It has been used extensively in the growing of tomatoes.

The basic principle of NFT is the simple one of circulating a shallow stream of nutrient solution over the roots of growing plants to provide adequate aeration, water and nutrients. The plants are grown in a series of sloping parallel gullies; the nutrient solution is pumped to the upper end of the gullies and flows along the lines of plants to a catchment tank sited below ground level.

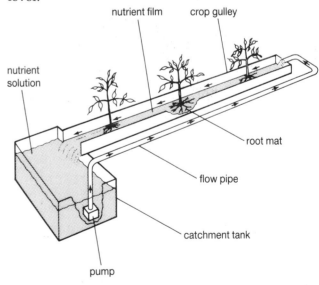

Figure 71   Nutrient film techniques

Adequate root aeration is ensured by circulation of nutrient solution in a film of water a few millimetres deep. If adequate aeration is not achieved the plants show symptoms of waterlogging and root death: in tomatoes, typical symptoms are a dark green appearance and the development of adventitious roots at the base of the stem.

An even slope must be prepared, free of depressions which would allow formation of deep pools of immobile solutions, within a gully and various methods are used to ensure an even slope. Gullies are usually formed from black or black and white polythene, which is cheap and has no stabilisers which can enter the solution and cause toxicity. Catchment tanks holding approximately 10% of the solution in circulation allow good mixing of nutrient solution and injected concentrated nutrient solutions, acid for pH control and fresh water to maintain the total volume of solution. A system supporting a mature tomato crop contains about 50 000 litres of nutrient solution per hectare. Automatic control of nutrient levels and pH is essential when tanks of this size are used because changes take place rapidly.

In areas such as the Channel Islands, where the presence of sodium in the water supply is a major problem, rain water collected from glasshouse roofs and stored is now commonly used for NFT.

> **Now try Investigation 19 Glasshouse Management in the *Plant Science in Action Investigation Pack*.**

8.13 ARTIFICIAL PROPAGATION TECHNIQUES

As seen in the previous chapters, propagation of crop plants by seed has considerable advantages. Each plant produces a very large number of seeds which are relatively easy to harvest, easy to handle and, with the exception of some tree seeds, they store well. In addition, they can be fed through machinery for mechanical sowing. There are however, a number of disadvantages, which have meant that, for certain groups of plant, vegetative or asexual propagation is preferable. Many methods of artificial vegetative propagation are used in the horticultural industry for a variety of reasons:

- The most widespread of the problems associated with propagation by seed is that sexual reproduction leads to variation. Increased mechanisation of the agricultural and horticultural industries, and the demands of the consumer, has increased the demand for uniform crops. Machines have been designed to cope with a standard plant and plants which differ from this standard will be missed, either not sprayed or not harvested, with a consequent loss of yield. Seed propagation will

only produce a uniform crop of standard plants if the crop is naturally inbreeding (such as cereals and legumes), or if it is possible to develop **F₁ hybrids**. However, many crop species do not lend themselves to either of these techniques. These species will only be acceptable to the mechanised industry if they can be propagated by vegetative methods, to produce clones of uniform plants. Vegetative propagation overcomes this problem by avoiding **meiosis** and fertilisation. New plants are produced entirely by **mitosis** and so have an identical genotype to their parent. All the desirable characteristics of the parent will be present in the offspring.

- In some species, especially woody plants and some bulb forms, the juvenile phase through which seedlings have to pass, may last for several years. Throughout this period the plant will only grow vegetatively – the juvenile phase being characterised by an inability to produce flowers. This is obviously a serious disadvantage if the plants are being grown for their flowers or fruit. It is uneconomic to tie up large areas of orchard with unproductive juvenile trees, or to have to grow daffodils for several years until the plants are mature and start to flower. By propagating from mature plant material, the juvenile phase may be avoided completely. The time between propagation and cropping can be greatly reduced and the problems of undesirable, juvenile features do not arise.

- Some crops are the products of complex breeding programmes involving **hybridisation**. Often the hybrid crop is sterile as with bananas, pineapples and seedless grapes. In these cases, although a fruit is formed, there are no seeds. This is both deliberate and desirable, since the fruit is used for eating and a fruit full of seeds is less attractive. However, this creates a problem with propagation. Either the hybridisation programme has to be repeated every year, a time consuming and elaborate process, or vegetative propagation techniques have to be employed.

When no natural system of vegetative reproduction exists, it is sometimes possible to initiate one artificially. This can be done in a number of ways as shown in Table 8.4.

Artificial propagation relies on the fact that plant cells possess the property of **totipotency**. This means that each cell contains all the genetic information necessary to produce a whole new plant. Plants are, therefore able to regenerate themselves from small pieces of tissue or individual cells.

Table 8.4   Artificial propagation

METHOD	PROPAGATION	CONDITIONS	MULTIPLICATION POTENTIAL	TIME
Cell division	shoots and roots		× 5	
Cuttings	shoots, leaves, roots.	prevent water loss	× 10–15	6–24 months
Grafting	scion to root	prevent water loss and keep warm		
Micropropagation	tissues, cells	aseptic conditions	× 10⁻⁶	3–5 years

8.14 CUTTINGS

More or less any part of a plant can be used as a cutting provided it contains cells which are able to reproduce themselves by division and then differentiate to form new tissue. Cells with this ability are found in the vascular cambium and the parenchymatous regions. This will generally be in and around the vascular bundles.

Cuttings are normally prepared from shoots but roots, leaves and even single buds may be used. In some cases, parts of perennating organs may also be used, although these are often formed from modified shoots. But in general, plant parts in the juvenile phase regenerate most easily and this ability decreases as the tissues mature. This decline in regenerative ability is associated with the gradual replacement of parenchyma tissue by woody tissues. However, even woody plant material can regenerate, provided it is prepared appropriately.

Cuttings must be kept alive and functioning while they replace those parts which are missing. They need to be protected from unfavourable environmental conditions and disease organisms, they need to be provided with those conditions which will enable them to form new tissue as quickly as possible. If the conditions can keep the cutting alive, long enough for it to reform the parts it needs to resume an independent existence, then the procedure will be successful.

8.14.1 Soft-tip cuttings – chrysanthemum

As a general rule, shoots in the **juvenile phase** will regenerate roots most readily. These non-woody, vegetative shoots are called **soft-tip cuttings** and the younger they are, the more readily they will regenerate roots.

Soft-tip cuttings are taken in the spring, as soon as there is sufficient growth to make a reasonable cutting. At this stage the shoots are growing fast and have a high rooting potential. But the young leaves have not developed their ability to control water loss. The cuticle in particular will be incomplete until the leaves are fully expanded and so the cuttings will be very susceptible to water loss.

Soft-tip cuttings

When the pots and cutting compost are ready for the next stage, the leaves are removed from the bottom third of each cutting and the stem is then cut immediately below a node. This provides a cut surface of hard, closely packed cells which will help prevent the spread of infections through the wound. To further reduce the risk of infection, the cutting base should then be dipped into a fungicidal powder. There is however no need for rooting hormones as soft-tip cuttings have a very high regenerative ability.

Cuttings are planted in cutting compost. A hole should be made in the compost with a dibber and the cutting placed in this hole, up to its leaves. It must not be pushed into the compost as this will damage the cut surface. Similarly, the cuttings should be firmed in by watering from above with a fine rose watering can, **not** by pressing on the soil.

A mist unit provides the ideal conditions for the rooting of such cuttings. The mist irrigation prevents water loss from the leaves by the maintenance of high humidity around the leaf and the leaves are kept cool by the latent heat of vaporisation, thus ensuring that shoot growth is kept to a minimum.

8.14.2 Hardwood cuttings

Hardwood cuttings are much more robust than soft-tip cuttings and so are easier to deal with. They may or may not possess leaves. Those without leaves can tolerate a much wider range of environmental conditions than soft-tip cuttings.

Hardwood cuttings use mature plant stems during their dormant period. Although most of the tissues are in the **mature phase**, there will be an area at the base of the stem which retains juvenility and it is from here that rooting will occur. However the rooting ability of mature tissue varies from species to species. Suitable species include blackcurrant, willow and gooseberry.

Cuttings are taken by cutting the stems at the base. The top of each stem is then removed by making a sloping cut immediately above the proposed top bud. It must be remembered that it is the basal region of the stem which has the greatest rooting potential. A horizontal cut is then made 15 cm below the top bud to produce a cutting 15 cm long. The base may then be dipped in rooting hormone.

Cuttings can then be planted out, no special protection being required. Well cultivated soil is important as a good oxygen supply is necessary for root formation. The cuttings should be buried so that only the top two or three buds protrude from the soil surface. The soil should then be well firmed around them. The cuttings can then be left alone throughout winter and the following growing season, during which time they will have formed two or three shoots and a root system. They can then be transplanted to their final positions in the autumn.

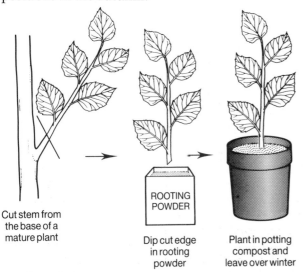

Cut stem from the base of a mature plant

Dip cut edge in rooting powder

Plant in potting compost and leave over winter

Hardwood cuttings

The regeneration of cuttings can be enhanced by the application of plant growth regulators on shoot and root growth. It is important to encourage the growth of roots initially so that the cutting can become established and obtain water and nutrients.

Table 8.5 The effect of auxin, cytokinin and gibberellin on cutting regeneration
* = P.G.R. affects growth

P.G.R.	LEVEL	ROOT GROWTH	SHOOT GROWTH
auxin	high	*	
	low		*
cytokinin	high		*
	low	*	
gibberellin	high		*
	low	*	

Synthetic hormones are usually used in rooting powders as they are less soluble than natural ones and so are less likely to be transported around the plant. It is important that the plant hormones stimulate root growth, and are transported to other parts of the plant, where they will have other physiological effects.

> **Now try Investigation 20 Rooting in Cuttings in the *Plant Science in Action Investigation Pack*.**

8.15 GRAFTING

Grafting is a technique for joining two pieces of living plant tissue in such a way that they will coalesce and subsequently grow and develop as one organism. It is a relatively difficult and time consuming process when compared with other methods of vegetative propagation and will usually only be used when a chosen **cultivar** proves too difficult to propagate by any other means.

8.15.1 Making the graft

Normally a shoot (**scion**) from one plant is grafted onto a root system (root stock) from another plant. To form an effective graft union, it is essential that the scion and root stock are married up in such a way that there is close contact between the vascular cambiums of scion and root stock. It is also necessary that they are brought together as quickly as possible. If the cut surfaces of either are allowed to dry, then the surface cells die off and a layer of dead cells separates the two cambial regions, preventing an effective union.

A graft union is achieved by the wound healing response in both the scion and the root stock. Initially a callus bridge is formed between scion and root stock. This results from the multiplication of cells in the cambial regions forming a mass of parenchymatous cells called a callus. The callus unites the scion and root stock forming a callus bridge and then within this a new cambium is formed, joining the cambium of the root stock and scion. Finally the new cambium produces new vascular tissue which links up with the existing vascular tissues and the graft is complete. It is however, important to note that this process does not involve mixing of cell contents and at all times the root stock and scion are separate genetic entities.

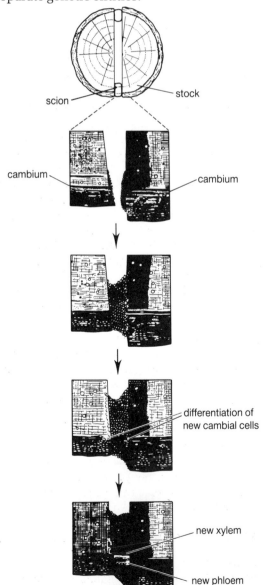

Figure 72 *Development of cambial cells during grafting*

8.15.2 Maintaining the graft

Once the graft has been made, it is very important to create an environment around the graft which encourages the graft union to form. This generally involves preventing water loss from the graft and if possible, keeping the tissue warm. Evaporation is prevented, either by enclosing the graft with an impermeable tape or by covering the whole area with grafting wax.

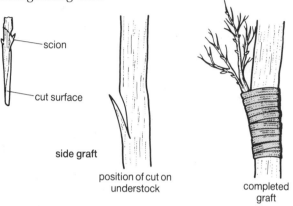

Figure 73 *Completed graft*

Once the union has formed, the success of the new plant will be much greater if competition from the root stock is prevented. Therefore all shoots from the root stock must be removed as soon as they appear. It is also a good idea, initially, to control the growth of the scion to prevent it becoming top heavy and tearing the graft off.

Grafting is a commonly used technique in the propagation of apples. Apple trees are hard to grow from seed because the juvenile phase lasts about five years and during this time no fruit is produced. Figure 76 shows the effects of grafting a scion from the same tree onto different root stocks.

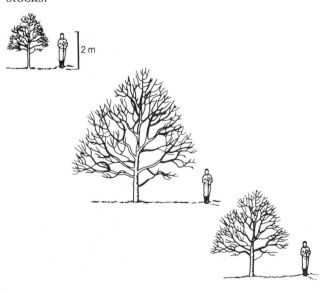

Figure 74 *The effects of grafting scions from the same source onto different root stocks*

In this example, the root stock seems to control the eventual size of the tree and the scion other characteristics such as size, shape, colour and yield of fruit. The use of grafting in apple trees is also useful as it allows the development of a range of trees each suited to a particular type of environment.

8.16 MICROPROPAGATION

Micropropagation involves the **clonal** (vegetative) propagation of crop plants on a very large scale and under carefully controlled conditions. By exploiting the inherent regenerative abilities of plants, it has made possible a revolution in plant propagation, and has overcome many of the barriers to understanding in other areas of plant biology. Vast numbers of identical copies of a plant can be produced very quickly and these can be screened for disease resistance, pesticide tolerance or tolerance of some environmental stress. It is even possible to evaluate a plant's ability to synthesise novel compounds which may have uses to mankind, either within the plant (as an insecticide for example) or in an extracted form (antibiotics, heart stimulants, fungicides). The number of plants produced is so large because they can be produced so quickly, so clonal propagation makes selective breeding programmes possible in a hitherto unimaginable scale.

The techniques used in micropropagation can be divided into those which use plant parts which already possess a shoot apical meristem (pre-existing shoot meristem) and those which use plant parts where there are no shoot meristems (no pre-existing shoot meristem). Commercial crop propagation tends to rely on the former technique: using stem sections with pre-existing shoot apical meristems. In principle this technique is very similar to taking cuttings. The main differences being that the plant parts used are very much smaller than a conventional cutting and the growing conditions are aseptic and carefully controlled.

There are four distinct stages which must be passed through if plants are to be propagated successfully, using micropropagation techniques. **Stage 1** involves the establishment of an aseptic culture; **Stage 2** is the multiplication stage, during which the pants regenerative abilities are exploited to increase the numbers of shoot apices from which new plants can be grown. During **Stage 3**, the shoot apices are grown into shoots and then rooted ready for planting into compost during **Stage 4** and their subsequent hardening off.

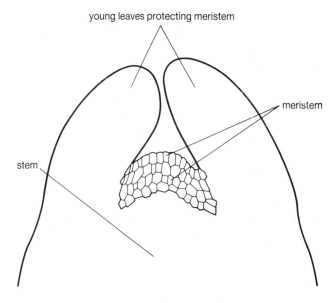

Figure 75   The tip of the shoot showing the meristem

8.16.1 Obtaining explants

Explants, taken from the mother plant are prepared by dissecting the plant stem so that there are a number of stem sections each with an apical meristem. The position of the meristem is shown in Figure 75. This is normally carried out in a laminar flow cabinet which provides a stream of finely filtered, spore-free air over the operators. The air current ensures that any potentially contaminating spores on the operators' clothing or hands are carried out of the cabinet, before they fall on to the plant tissues. The surfaces of the cabinet and the dissecting instruments are also sterilised regularly during the operation. The explants themselves are surface sterilised by washing in a dilute hypochlorite solution and they are then placed on a sterile culture medium in a glass jar or tube. The culture medium contains a number of chemicals, but most importantly, a high concentration (20–30 g/l) of sugar (usually sucrose). One of the interesting features of micropropagation is that the explants are able to adopt a heterotrophic method of nutrition provided they are supplied with a source of carbohydrate. The culture medium will also contain a range of mineral nutrients, vitamins and plant growth regulators in an agar base. The agar is important to provide support for the developing shoot and to prevent the tissues from sinking into the culture medium where the lack of oxygen would be a problem.

8.16.2 Growing the explants

The explants, on their culture medium in sealed glass jars or tubes are then placed in a growth room where temperature and light are controlled so that they start to grow (Figure 76).

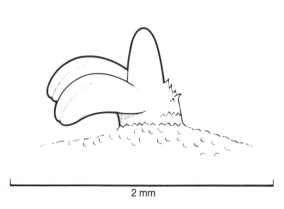

Figure 76   Cells of the explant sealed in culture medium begin to multiply.

In an intact plant, shoot growth is promoted by cytokinins from the roots. However the explants have no roots so it is necessary to provide the 'hormonal signal' from the culture medium. At the levels one might expect to find in a plant, cytokinins cause the production of a single, straight shoot. However, for the purposes of multiplication, it is necessary to suppress elongation of the main shoot and stimulate shoot formation by the axillary buds. The development of several axillary buds on each shoot gives rise to the multiplication and is achieved by increasing the cytokinin concentration of the culture medium; high levels of cytokinin produce a short but many branched shoot culture. At regular intervals, usually about a month, these branched cultures can be divided by cutting off the branches and growing each of them on fresh culture medium, also with a high cyctokinin level, so that the axillary buds form branches. This divide-and-multiply process can be repeated at monthly intervals, more or less indefinitely, producing exponential growth in the numbers of shoots. The number of axillary shoots per stem varies with species but with a multiplication factor of four, that is, for each shoot producing four axillary shoots per month, there will be over *16 million* shoots available for rooting at the end of the year!

8.16.3 Growing on

When sufficient shoots have been produced, they are recultured onto media containing lower cytokinin concentrations, more similar to the concentrations normally found in plants. At these concentrations of cytokinin, shoots grow tall, and axillary buds remain dormant and eventually the shoots are large enough to be used as 'mini-cuttings' in the rooting phase.

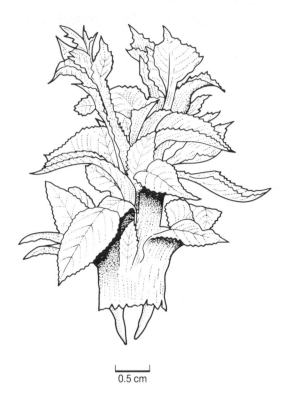

Figure 77   Mini cuttings ready for rooting

8.16.4 Hardening off

The transition from the protected environment of a culture jar to the much more hostile environment in a glasshouse is a critical phase. In the culture jar the shoots have been feeding heterotrophically, they have been protected from water stress and have been kept in constant light and temperature regimes. They tend to have poorly developed waxy cuticles and their stomata are fixed in the wide open position. They need to be 'weaned' and this takes time. They need to prepare for autotrophic nutrition, to gain control over their stomata and to produce leaves with a waxy cuticle.

Weaning is achieved by placing plants in a fogging unit. This creates a fine mist of tiny water droplets, similar to those found in clouds, so that the plants can develop in a water saturated atmosphere where the relative humidity is 100%, without suffering water stress. The unit also gives partial shade. There is a danger of photo-oxidation of photosynthetic pigments in full sunlight and the plants need time to adapt to the much brighter daylight conditions. Temperature fluctuations will also be controlled as far as is possible.

> **Now try Investigation 21 Artificial Propagation of Plants in the *Plant Science in Action Investigation Pack*.**

QUESTIONS

1 The atmosphere within a greenhouse is often enriched with carbon dioxide.
   a) How is this done?
   b) What effect will enrichment have on the plants?
   c) Why is the level of carbon dioxide only raised to 0.09%?
   d) Why is enrichment only carried out at night?

2 a) List the main problems associated with the supply of water and nutrients to plants growing in soil in a greenhouse.
   b) How are these problem overcome by the use of NFT?

3 Chrysanthemums are short day plants.
   a) What does this mean?
   b) How are growers able to manipulate the environment so that flowering chrysanthemums are produced all year round?

4 a) It is advisable to take soft-tip cuttings when the cells are turgid. Why?
   b) Look at the graph below:

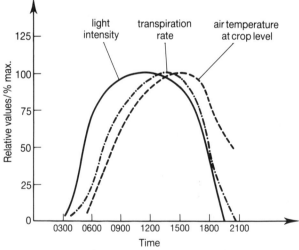

'The relationship between light intensity, air temperature and transpiration rate in lucerne leaves'

   (i) Explain the effect that increasing temperature and light intensity have on the rate of transpiration.
   (ii) Why is it an advantage to take cutings during the early morning?
   (iii) Why are soft-tip cuttings taken in the afternoon less likely to survive?
   (iv) What precautions should be taken to prevent water loss once the cuttings have been taken?

5 Apples are self-infertile. This means that they cannot transfer pollen from anther to stigma of the same plant, so at least two varieties of apple must be in close proximity to each other if pollination and thus fruit production is to occur. How might grafting help overcome this problem?

6 Micropropagation has become increasingly popular for the production of many types of plant.
   a) List the advantages of micropropagation over the other types of vegetative reproduction discussed.
   b) What are the main problems associated with this technique?
   c) Use the data shown in Table to draw a graph showing the impact of micropropagation on plant breeding.

Numbers of plants produced by micropropagation in the Netherlands

NUMBERS OF PLANTS	1984 $\times 10^4$	1985 $\times 10^4$	1986 $\times 10^4$
Pot Plants	1500	1700	2000
Cut Flowers	1000	1100	1300
Agricultural Crops	30	30	30
Vegetables	6	7	7
TOTAL	2536	2837	3337

BIBLIOGRAPHY

Butcher, D.N., Ingram, D.S. *Plant Tissue Culture. (IOB)* Edward Arnold.

Larson, R.A. (1980) *Introducton to Floriculture.* Academic Press.

Janick, J. (1986) *Horticultural Science.* W.H. Freeman.

Sandiforth, A. *Cloning The Bramley. (SATIS 16–19 unit 37).* ASE.

Schery. R.W. (1972) *Plants for Man.* Prentice Hall.

9 BREEDING CROP PLANTS

LEARNING OUTCOMES

After studying this chapter you should be able to:

- distinguish between discontinuous and continuous variation,
- contruct and interpret frequency distribution curves,
- explain the contributions made to variation by the environment and genetics,
- recall the fact that characteristics are inherited from one generation to the next through the passing on of genes which are found on the chromosomes,
- recall that during sexual reproduction chromosomes are randomly distributed within the gamete cells,
- state the 'Laws of Inheritance' and explain how characteristics are normally passed on from one generation to the next,
- distinguish between cases involving co-dominance and those showing normal dominance patterns,
- define the terms 'multiple genes' and 'polygenes' and describe the difference between them,
- outline the process of epistasis,
- define the term 'heritability',
- calculate the heritability of characteristics from data which is supplied, and comment on the significance of the results obtained,
- distinguish between 'inbreeding' and 'outbreeding',
- discuss the methods used by breeders to produce better crop plants.

9.1 TYPES OF VARIATION

Throughout the world a very large proportion of agricultural and horticultural crops are grown directly from seed in fields. Because of this it is important that all aspects of crop development, from seed germination right through to readiness to harvest, are synchronised. Every plant within the crop must reach a similar developmental stage at the same time. However, achieving this uniformity requires careful control of their life cycle.

Individuals within a population vary from each other. Variation may be **discontinuous** where the characteristics exhibited fall into a few sharply defined groups, or **continuous** where a whole range of differences are possible falling between two extremes. Most individuals exhibit characteristics with values approaching the mean, giving rise to a normal distribution throughout the population as a whole.

Continuous variation is far more common. In examples of continuous variation, a whole range of intermediates are possible ranging between the two extremes. Figure 78 shows an example of continuous variation.

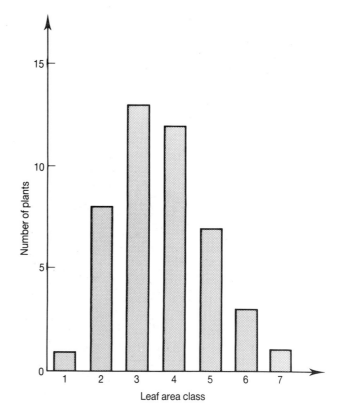

Figure 78 Frequency distribution of leaf area in swedes

The leaf area of 45 swede plants is measured after seven weeks. The areas ranged from 187 cm^3 to 1216 cm^3. The graph clearly shows how the leaf areas range between these two extremes, with the majority of the plants having leaf areas of around the mean.

> **Now try Investigation 22 Continuous Variation in the *Plant Science in Action Investigation Pack.***

9.2 THE CAUSES OF NATURAL VARIATION

Plants within a crop will vary from each other naturally. To ensure that a uniform crop is produced, scientist and grower must select the most suitable husbandry techniques and crop characteristics.

This variation is due to a combination of two components:

(i) **The environmental component** which may include influences such as differences in soil type within a field, attack by pests and the interaction of weeds with crops.

(ii) **The genetic component** which is due to inherited differences.

Variation within a crop which is due to the environment (such as the effects of rust infection on the yield of kernels in wheat as shown in Figure 79) can be limited by good husbandry techniques. Genetic variation relies on the selection of specific inherited characteristics which can be incorporated into a breeding programme to produce standard seed of a high quality.

9.3 THE GENETIC COMPONENT OF VARIATION

A proportion of the variation observed between two plants of the same species is due to genetics. Characteristics are inherited from one germination to the next because they are controlled by specific segments of the DNA molecule which makes up a large part of the chromosome found in the nucleus of a cell. These segments are called genes. Thus one characteristic is controlled by one or more pairs of genes.

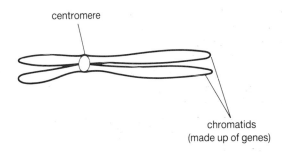

Figure 80   The chromosome

9.3.1 Chromosomes and the Gamete Cells

Every plant and animal is made up of a collection of body (somatic) cells and reproductive (gamete) cells. Each characteristic is controlled by one or more pairs of genes which are arranged on homologous pairs of chromosomes. The somatic cells each contain a complete set of homologous chromosome pairs for each characteristic the individual expresses. These are known as diploid cells.

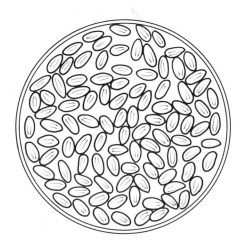
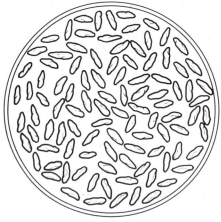

Figure 79   Variation in kernels obtained from a healthy and rust infected wheat plant

The gamete cells are formed by meiosis cell division. During meiosis the cell divides twice.

- During the first division, the chromosomes double their number and are equally divided between the two cells which form.
- During the second division, the chromosomes do not double. The homologous pairs split and one of each pair of chromosomes enters each of the two cells formed.

The resulting cells contain only half the original number of chromosomes. These are called haploid cells. Meiosis encourages variation as the chromosomes enter the haploid gametes completely at random. This is summarised in Figure 81.

During sexual reproduction, two gamete cells will combine to form a zygote. This cell, which will eventually form a new individual will, therefore contain genetic information from each parent. The new combination of genes will cause the offspring to be genetically different from each parent but still exhibit some similar characteristics.

Figure 81   An outline of gamete formation by meiosis

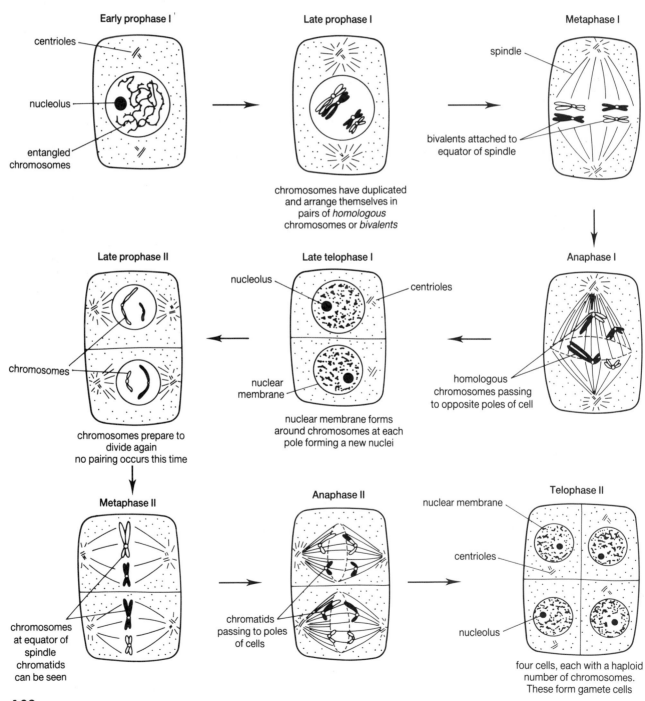

Early prophase I
- centrioles
- nucleolus
- entangled chromosomes

Late prophase I

chromosomes have duplicated and arrange themselves in pairs of *homologous* chromosomes or *bivalents*

Metaphase I
- spindle
- bivalents attached to equator of spindle

Anaphase I
- homologous chromosomes passing to opposite poles of cell

Late telophase I
- nucleolus
- centrioles
- nuclear membrane

nuclear membrane forms around chromosomes at each pole forming a new nuclei

Late prophase II
- chromosomes

chromosomes prepare to divide again no pairing occurs this time

Metaphase II
- chromosomes at equator of spindle chromatids can be seen

Anaphase II
- chromatids passing to poles of cells

Telophase II
- nuclear membrane
- centrioles
- nucleolus

four cells, each with a haploid number of chromosomes. These form gamete cells

9.4 GENE INTERACTION

In 1865 Gregor Mendel completed his now famous work on inheritance. An Austrian monk, Mendel designed experiments which followed the inheritance of well defined characteristics in the garden pea *Pisum sativum*. His findings led to the formulation of the two Laws of Inheritance which form the basis of modern day genetics. Mendel's work is well documented in A level texts and so will be dealt with only briefly here. (See especially *Genetics and Evolution* by Michael Carter in the **Focus on Biology** series.)

Pisum sativum

9.4.1 The Laws of inheritance

The First Law of Inheritance: The Law of Segregation.

'The characteristics of an organism are determined by internal factors (genes) which occur in pairs. Only one pair of such factors can be represented in a single gamete.'

The Second Law of Inheritance: The Law of Independent Assortment.

'Each member of a gene pair may combine randomly with either of another pair.'

Mendel discovered that each characteristic is determined by a pair of alleles or genes. Each gene has two forms – one dominant to the other – and this form is always expressed if it is contained within the genotype (the genetic make-up) of an individual's cells. The other form of the gene is said to be recessive.

For example, *P. sativum* produces two types of seed – one round and one wrinkled. The characteristics of seed appearance is controlled by a pair of genes which can exist in two forms:

(i) Dominant form (represented by R) and producing round seeds.

(ii) Recessive form (represented by r) and producing wrinkled seeds.

Each pea plant must contain a combination of two of these gene forms (RR, Rr, rr).

When gametes are formed, each will contain one of these genes determining seed appearance. During sexual reproduction, two gametes fuse to form a zygote. This new cell will then contain two genes determining seed appearance – one from

If R represents the gene for round seeds
 r represents the gene for wrinkled seeds

Each individual must contain a pair of genes for seed shape.

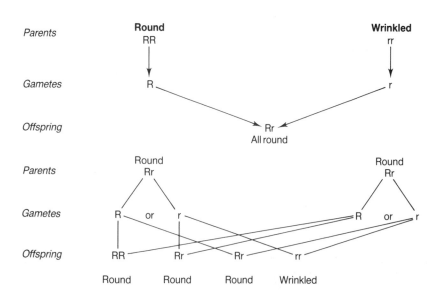

Figure 82 *Inheritance of seed appearance in peas*

each gamete. This new pair of genes will determine the appearance of the seeds on the new plant which develops from the zygote. Thus the actual observed characteristics that an individual possesses (the phenotype) will depend on the interaction of genes contained within the gametes from which they form. Some examples are shown in Figure 82.

9.4.2 Co-dominance

Not all pairs of major genes exhibit full dominance. The flower colour in *Antirrhinum* (snap dragon) is controlled by two genes R (red) and r (white). Inheritance is similar to that for Mendel's peas, except that R is not dominant to r. If an individual contains both gene forms within its genotype (Rr), pink flowers are produced.

Snap dragon flower colour is controlled by co-dominance.

9.4.3 Multiple genes

Some characteristics are controlled by more than two gene pairs. If a characteristic varies continuously throughout a population, such as seed weight, it is usually controlled by a large number of genes known as **polygenes**. These will be discussed later in this unit. If, however, the characteristic varies discontinuously and produces distinct groups of individuals, it is likely to be controlled by **multiple genes**. Multiple

genes sometimes have a cumulative effect on the phenotype, and sometimes show dominance, but not always.

In 1909 Nilsson Ehle carried out work on the wheat species *Triticum aestivum*. This species produced kernels which are either red or white. When crossing a pure breeding red plant with a pure pure breeding white plant, he noticed the first generation all produced red kernels as expected, and that in most cases, the second generation individuals were produced according to the expected ratio. However, some families produced second generation plants that were *all* red. If these red plants were allowed to interbred, the resulting generation gave a series of different ratios of red to white plants: 3:1, 15:1, 63:1, and all red.

It was concluded that kernel colour in this species was determined by three independent gene pairs (multiple genes). The dominant genes (R_1, R_2, R_3) giving red kernels, and the recessive genes (r_1, r_2, r_3) giving white kernels. If any of the three gene pairs contained the dominant allele, the kernel would appear red.

9.4.4 Polygenes

Many characteristics are controlled by many pairs of genes rather than one single pair. **Polygenes** are collections of large numbers of genes, each of which are individually of relatively minor importance but which collectively have cumulative importance on the value for one particular trait. They are additive in their effects, and so control examples of continuous variation.

Height in plants is an example of continuous variation. Within a population, plant height will vary over a range, with most individual's height falling around the mean. Height is an example of polygenetic inheritance.

Let us assume that height is controlled by three pairs of genes, Aa, Bb, Cc. The inheritance of the dominant gene allows for two height units, whereas the inheritance of the recessive form allows for the inheritance of one unit.

Thus, the tallest individuals in a population will have a genotype of AABBCC and will be 12 units tall, whilst the smallest individual will have the genotype aabbcc and will be only 6 units tall. A number of intermediate genotypes can exist between these two extremes:

$$AaBbCc = 9 \text{ units}$$
$$AAbbCC = 10 \text{ units}$$
$$AaBBCC = 11 \text{ units}$$

Each produces individuals of heights which fall between 6 and 12 units. If these are plotted onto a histogram, a normal distribution will be produced as shown in Figure 83.

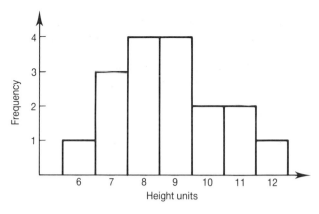

Figure 83 Frequency of height in plants are controlled by polygenes

9.4.5 Epistasis

In the examples considered so far, characteristics have been inherited from one generation to the next because the gene controlling each characteristic has existed as one of a pair. Each pair has a dominant and a recessive form, and the dominant form is always expressed when present. However, in some cases of inheritance, the expression of one gene is controlled by the presence or absence of another gene which may even exist on a different chromosome. This is called **epistasis**.

Onion colour is controlled by a number of pairs of genes which interact with each other. For ease we shall concentrate on the interaction of two of these pairs. Onions can be red skinned, yellow skinned or white skinned.

Red and yellow skins are controlled by a pair of genes:

R is dominant and gives a red skin

r is recessive and gives a yellow skin.

For either of these colours to be expressed the dominant form of the 'colour gene' (C) must also be present in the genotype. The recessive form of this gene (c) will inhibit colour production and result in white skins. This colour gene is called an epistatic gene as it controls the expression of another gene. Table 9.1 shows all the possible genotypes for each phenotype (skin colour).

Table 9.1 Onion skin colour: genotypes and phenotypes

	POSSIBLE GENOTYPES	POSSIBLE PHENOTYPES
Red	RRCC, RrCC, RRCc, RrCc	Red skin
Yellow	rrCC, rrCc	Yellow skin
White	RRcc, Rrcc, rrcc	White skin

9.5 HERITABILITY

The previous example shows how an individual's characteristics are influenced by the genes that they inherit from their parents and by the environment in which they live (their diet, climate, etc). When breeding animals and plants in order to attain a particular characteristic, such as high body mass, breeders need to know what proportion of the variation in that characteristic is due to inheritance. If inheritance has a large effect on the characteristic then mating two animals, both with high body mass, should produce offspring also with a high body mass. If inheritance has only a small effect, the breeder needs to concentrate on aspects of the environment. The genetic or inherited influence of a characteristic is called **heritability**.

9.5.1 Calculating heritability

In 1913 Emerson and East investigated inheritance of ear length in maize (*Zea mays*). They developed two pure breeding strains of maize, one with long ears (13–21 cm), and one with short ears (5–8 cm). Individuals from each strain were crossed to produce offspring (the F_1 generation) and individuals from this generation were crossed to give a second (F_2) generation. In each generation the ear length of the maize was measured.

Frequency distribution

Table 9.2 below shows ear length of the maize used in this experiment:

Table 9.2

Ear length (cm)	5	6	7	8	9	10	11	12	13	14	15	16	17	18	19	20	21
Parents SP	4	21	24	8													
LP									3	11	12	15	26	15	10	7	2
Offspring F_1					1	12	12	14	17	9	4						
F_2			1	10	19	26	47	73	68	68	39	25	15	9	1		

(after Emerson and East, from *An Introduction to Genetics* by A. Sturtevant and G. Beadle, W. B. Sanders and Co, 1941)

SP = short-eared parent   LP = long-eared parent F_1 = first generation F_2 = second generation

These results can be shown on a histogram. Variation in ear length in maize is an example of continuous variation.

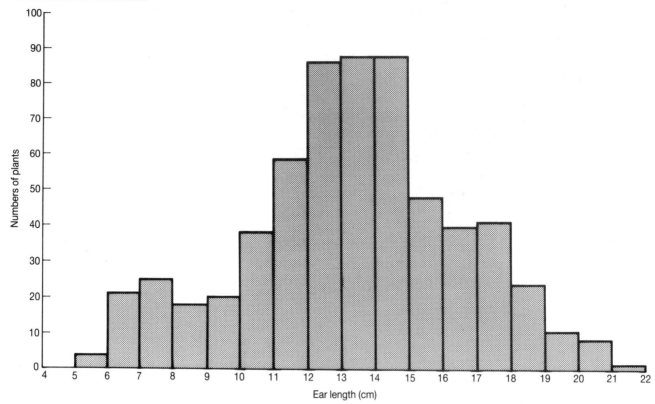

Figure 84 *Histogram showing variation in ear length in maize*

9.5.2 Means and variance

Variance describes the way in which data varies around the average value or mean. The histograms produced show that most individuals in a generation have a particular ear length (the modal class), with others showing some variation around this class. As you know variation has two components: genetic or inherited, and environmental.

Thus:

$$\begin{array}{ccccc} \text{Total} & = & \text{Genetic} & + & \text{Environmental} \\ \text{Variance} & & \text{Variance} & & \text{Variance} \\ \text{(Vt)} & = & \text{(Vg)} & + & \text{(Ve)} \end{array}$$

Variance is calculated using the following equation:

$$\text{variance} = \frac{\sum n\,(x - \bar{x})^2}{\sum n - 1}$$

Equation for variance

$\sum$ = the sum of
n = the number of individuals in sample
$\bar{x}$ = mean
$x - \bar{x}$ = difference between individual and mean

9.5.3 Heritability

Heritability is the proportion of total variance which is due to inheritance or genetics and not the environment.

$$\text{Heritability (H)} = \frac{Vg}{Vt}$$

Heritability should range between 0 and 1. It can also be expressed as a percentage. If it is calculated to be above 50% this indicates that inheritance plays a large part in the expression of the tested characteristic.

9.6 WHY DO WE NEED BETTER PLANTS?

Humans have been trying to improve the plants that they grow for food for thousands of years. The domestication of crops began over 8000 years ago, and ever since growers and scientists have been working to create new strains which yield more food of a higher nutritional value, use water and light and minerals more efficiently, are easier to harvest, and are resistant to disease. Many techniques have been developed, including hybridisation, breeding inbred lines and polyploidy. These techniques all rely on a knowledge of genetics and the natural breeding systems used by the plant.

9.6.1 Inbreeding and outbreeding

Plants generally fall into two groups: the self-pollinated **inbreeders**, which exhibit very little variation and the cross-pollinated **outbreeders**, which are much more variable.

Inbreeding has the effect of maintaining a largely uniform population, with variation arising from chance mutations. Inbreeding is advantageous to plants which occupy relatively stable environments, but because it generates much less variation than outbreeding, it is less significant from a long term evolutionary point of view.

Outbreeding is important in plant breeding as it allows characteristics from two plants to be brought together into a single variety but it creates problems when a chosen variety needs to be propagated.

A great many crops readily self-pollinate. Such plants give rise to **pure lines**, the members of which are descended from a single plant and are highly uniform. Fortunately, plants which inbreed naturally, do not show the loss of vigour sometimes known as inbreeding depression which is normally associated with enforced inbreeding. Most cereals and many members of the *Leguminosae* are self-pollinators and so their propagation is relatively straight forward.

Now try Investigation 23 Comparing Pollination in Outbreeders and Inbreeders and Investigation 24 Preventing Self Pollination in Inbreeders in the *Plant Science in Action Investigation Pack*.

9.6.2 Mechanisms to ensure outbreeding

It is the natural outbreeders, with mechanisms to prevent self-pollination, which present problems in seed propagation. In such species, self-pollination is prevented by a variety of systems. These either attempt to prevent pollen transfer from anthers to stigma within the same flower – **dichogamy** – or ensure that only pollen from another plant can carry out fertilisation – **incompatibility**.

9.6.3 Dichogamy

Dichogamy means that the anthers and the stigma mature at different times so that the anthers are releasing their pollen either before or after the stigma is receptive. The most common situation is for the anthers to ripen first (**protandry**, such as in willowherb, geraniums, wood sage) but it is much more effective if the stigma ripens first (**protogyny**, such as in bluebells, ribwort, plantain).

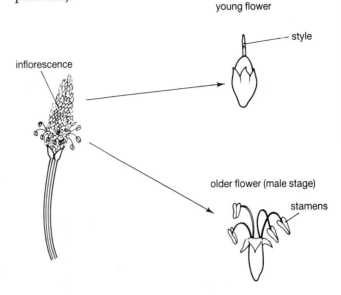

Figure 85   Protogyny in ribwort plantain

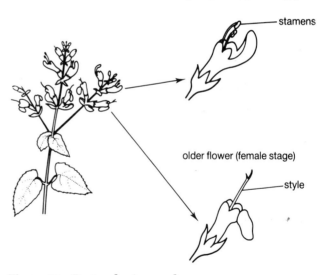

Figure 86   Protandry in wood sage

9.6.4 Incompatibility

Incompatibility mechanisms are much more effective at preventing self-pollination as they apply not only to all the flowers on the same plant but to flowers on cloned plants too. Plants are able to discriminate against pollen grains which possess alleles that are the same as their own genotype. There are two types of incompatibility: **sporophytic** such as in *Brassica sp.* in which the pollen fails to produce a pollen tube on the stigma; and **gametophytic** as found in *Nicotiana sp.* in which the pollen tube is 'blocked' in the style.

It has been established that incompatibility usually depends on a single genetic locus known as the 'S gene'. In a population, this can occur in a large number of different forms or alleles. Any one diploid plant will have two of these alleles: S_2S_6 whereas its pollen will only have one, *either* S_2 *or* S_6. The protein products of the S gene are carried by the pollen and can be recognised by the parent plant. This will respond in such a way that, either the potentially self-fertilising pollen grain is prevented from forming an effective pollen tube, (sporophytic incompatibility) or the 'self' pollen tubes wither before reaching the ovary, (gametophytic incompatibility). Non-self pollen carries a different incompatibility protein and is not inhibited in this way.

Now try Investigation 25 Investigating Methods of Overcoming Self Incompatibility in the *Plant Science in Action Investigation Pack*.

9.7 PRODUCING BETTER CROP PLANTS

Inbreeding is advantageous to breeders as it produces genetically stable seed which decreases the likelihood of variation within the growing crop. However many crop plants possess mechanisms to prevent inbreeding as outbreeding enables them to survive better within a natural environment as it encourages variation. Faced with these defences against inbreeding, seed producers have had to develop breeding strategies to overcome these problems if they are to produce genetically standard seeds.

9.7.1 Producing 'pure lines' by selection

Even before humans began to domesticate selected crop plants, species were subjected to natural selection. Varieties of plants which were well suited to the environment in which they lived, thrived, whereas those which were unable to compete failed and eventually died out.

Humans started to domesticate crop plants such as flax and barley, over 8000 years ago in the Middle East. As the domestication of plants spread throughout the world, new varieties developed naturally which were able to grow in the particular environment in which they found themselves.

Soon humans began to improve varieties of crop plants which had developed as a result of natural selection. Early farmers selected traits which were desirable and saved the seed from the plants which showed this particular trait. The selected

seeds were then planted and the resulting plants were allowed to cross pollinate with others showing the trait. The 'best' seeds were selected again and replanted – this process being repeated for several generations. This method eventually gave rise to a population all of which showed the desired trait. Pure line selection is illustrated in Figure 87.

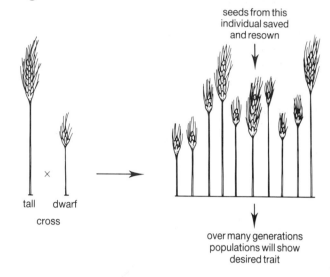

Figure 87   Pure line selection

A modern example of pure line selection is the breeding of corn to increase its yield of protein and oil. Figure 89 shows how over successive generations it is possible to increase the percentage of both oil and protein the corn contains by selecting seed from high yielding plants. After fifty generations this variety of corn has a high oil and protein content.

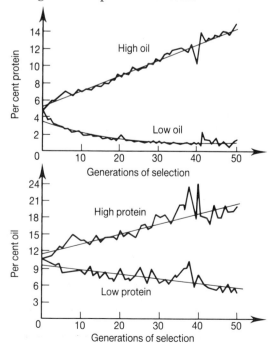

Figure 88   The effects of pure line selection on oil and protein content of corn

9.7.2 Inbreeding depression

The main disadvantage of producing pure lines in seed production is that often the species will suffer from **inbreeding depression**. Inbreeding depression is often characterised by lack of resistance to disease, low germination rate and general lack of vigour in the plant. This will lead to reductions in yield.

It is a feature of pure line selection that the resulting plants are largely **homozygous**. This means that most of their characteristics are controlled by two genes e.g. tt or TT. Characteristics which are controlled by one of each gene form e.g. Tt, are said to be **heterozygous**. It is thought that as plants are bred to become homozygous for one desirable characteristic they also become homozygous for other, less desirable, characteristics. These are usually controlled by recessive genes and so would be 'masked' by the dominant gene in heterozygous individuals.

9.7.3 Producing hybrids

Hybrid seeds do not show inbreeding depression and are usually more hardy than normal seed. Hybrids can be produced by selecting two homozygous inbred lines which possess the desired traits. Self-pollination must be prevented, especially in the line that is to be used for seed production (the female line). This can be done by:
  (i) using species in which the males are sterile
  (ii) using species which exhibit dichogamy
  (iii) emasculation – the removal of the male flowers or parts of flowers on the female line plants, before they develop. This is difficult to do on a large scale unless the species possesses separate male and female flowers (e.g. maize).

Once self-pollination has been prevented, cross-pollination must be encouraged and the seeds that form on the 'female' plant are then collected. These seeds are called F_1 **hybrids** and contain a mixture of desirable characteristics.

Table 9.3 shows how F_1 hybrids are more vigorous than normal seed and seed produced by pure line selection.

Table 9.3   Hybrid wheat yields

SOURCE OF SEED	AVERAGE YIELD t/Ha
Normal	6.00
Inbred after 30 generations	1.50
Hybrid of inbred	5.00
Offspring of hybrid	2.50

F_1 hybrids are fairly difficult to produce and, because they are highly heterozygous, the seeds that the plants produced cannot be saved and resown. This would produce a very variable crop. This means that growers have to buy new seed each year.

From a commercial point of view, maize is the most important crop plant to be propagated by the production of F_1 hybrids. Maize is **monoecious** (i.e. the male and female flowers although on the same plant are well separated), and because of this, it is relatively easy to produce all female plants by removing the terminal male flowers. It is therefore very easy to control pollination.

Some *Crucifers*, such as Brussels sprouts, have a incompatibility system which prevents self-fertilisation. However, by artificially pollinating **young** flowers which have not quite matured and do not yet have a functional incompatibility system, it is possible to force self-pollination to produce inbred lines. Two such inbred lines, carrying different incompability alleles can then be allowed to intercross naturally to produce F_1 hybrid seed. Self-pollination is prevented by the incompatibility system.

9.7.4 Polyploidy

As you are aware somatic (body) cells contain a **diploid** (2n) number of chromosomes and gamete (reproductive) cells contain a **haploid** (n) number. It is possible to produce individuals with multiple numbers of chromosomes such as triploids (3n), tetraploids (4n) etc. These **polyploids** often exhibit desirable characteristics like disease resistance or increased vigour. It is estimated that more than 100 000 species of angiosperms are polyploids.

Triticum aestivum a wheat whose flour is used in bread production, has been selectively bred by crossing wild wheat and grass varieties, as shown in Figure 89.

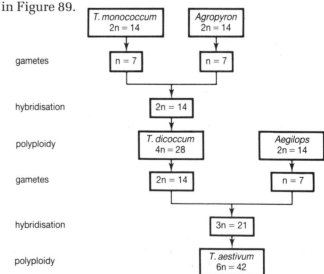

Figure 89   *The breeding of bread wheat*

T. monococcum is unsuitable for domestication because its ears are very brittle when ripe and it is fairly low yielding. This variety was crossed with a grass *Agropyron* producing a hybrid wheat (2n = 14) which underwent polyploidy to produce *M. dicoccum* (2n = 28). This was crossed with another grass species (*Aegilops* – 2n = 14) to produce a triploid hybrid. After polyploidy this produced *T. aestivum* (6n = 42).

Polyploidy can occur naturally or can be induced by the use of chemicals such as colchicum. However, sometimes polyploid species are sterile and unable to form gametes.

9.8 GENE TRANSFER IN PLANTS

9.8.1 What is gene transfer?

Chemically, DNA is a relatively simple polymer made up of only four different base units linked together to produce 'genes'. The basic structure of all genes is therefore fairly similar. It is therefore, possible in some cases, to isolate genes which control favourable characteristics in a particular organism, and transfer them to cells taken from a different organism. These genes are then incorporated into the normal DNA of that cell and function as part of that DNA.

9.8.2 Gene transfer – procedure

The procedure of gene transfer involves four distinct stages.

(i) Isolation and slicing of desired DNA segment

The DNA molecule is extremely long and fragile. Extracted DNA is treated with **restriction endonuclease** enzymes. These recognise particular sequences of bases within the DNA strand and cut the DNA into short segments between identified base pairs. DNA is made up of two strands: cutting sometimes occurs at the same position on each strand leading to the formation of 'blunt ends' or it occurs in a staggered fashion, forming 'sticky ends'. These DNA fragments are often called **passenger DNA**.

(ii) The insertion of passenger DNA into a vector

If the fragments of DNA produced are to be introduced into a host cell, they must first be incorporated into a **vector**. A vector is a piece of DNA which can combine with the passenger DNA to form what is known as **recombinant DNA**. It is

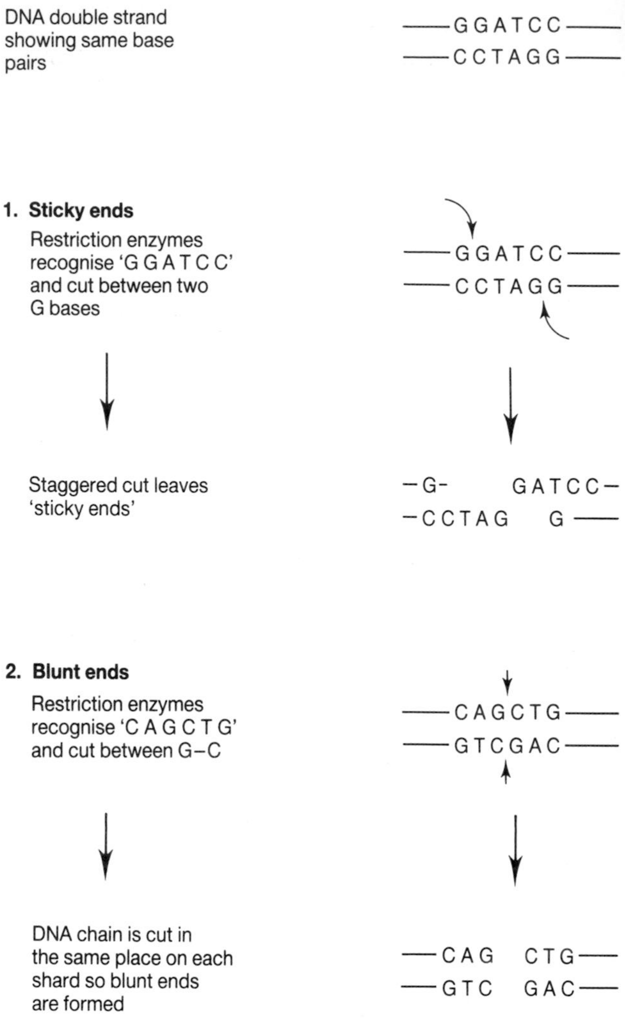

Figure 90   The formation of blunt and sticky ends when DNA is treated with restriction enzymes

the recombinant DNA which is introduced into the host cell. The main function of the vector is to allow the DNA to enter the host cell. The vector has sections which are recognised by the host cell and so the cell will allow its entry and combination with its own DNA. The attached passenger DNA is also incorporated.

Passenger DNA joins with the vector DNA due to the action of **DNA ligase** enzymes. These enzymes recognise the blunt or sticky ends produced by the restriction enzymes. If complementary sequences are found within the vector DNA, the ligase enzyme will cause the passenger DNA to be incorporated into the vector DNA.

There are two types of vector:
● **Plasmid vectors** are free, circular pieces of DNA found in the cytoplasm of bacteria. These can be extracted in the laboratory by the breakdown of the bacterial cell wall, and centrifugation. Plasmid vectors are widely used to introduce break down **foreign** genes into bacterial cells. This technique is used in the production of bacterial products, such as insulin.

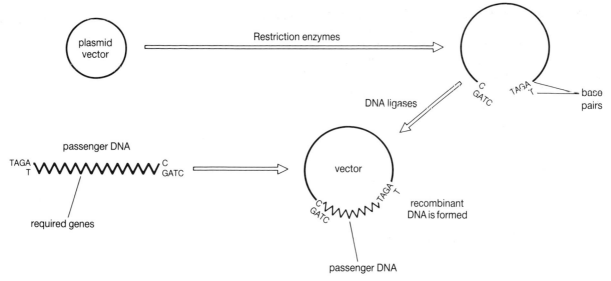

Figure 91 Plasmid vectors

• **Virus vectors** The normal virus life cycle has been exploited as a method of transferring genes. Viruses can only survive inside living cells. Once a cell is infected, the virus releases its DNA and this is incorporated into the DNA of the cell. As the cell replicates, so does the viral DNA. If desired genes are introduced into a viral carrier, these genes will also be replicated.

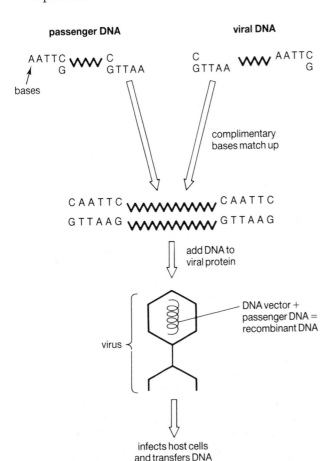

Figure 92 Virus vectors

(iii) Selecting recombinant DNA fragments

During gene transfer it is not possible to control the progress of the reactions described in (i) and (ii) so that only recombinant DNA containing desired passenger DNA is produced. As many different fragments of DNA are produced during the slicing process, many recombinant DNA sections are formed. In order to select the desired recombinant DNA, one of the following procedures is required:

• **Radioactive labelling** Many chemical elements have radioactive isotopes which can be detected by using a Geiger counter, and so these isotopes can be used to trace the pathway of chemical elements in metabolic processes. This technique can be used to identify the presence of particular base pairs within a DNA fragment and so be used to identify desired recombinant DNA sequences.

• **Marker rescue** If one of the genes within the recombinant DNA is responsible for the synthesis of a particular product, the presence of this product or marker can be detected when recombinant DNA is supplied with the substrates required for its formation.

(iv) Introduction of recombinant DNA

In most cases of gene transfer, recombinant DNA is injected either as a plasmid or a virus, into a bacterial host cell. These cells will incorporate the 'foreign' DNA and rapidly reproduce it. One of the main uses of gene transfer commercially, is the production of biological chemicals such as antibiotics, rennet and insulin. Bacterial cells containing the appropriate DNA, grown in culture and supplied with the required substrate, will synthesis these products much more rapidly than their animal equivalents. It is, however, possible to introduce recombinant DNA directly into the cells of higher organisms. This is discussed later.

9.8.3 The use of gene transfer in crop production

The technique of gene transfer is of great importance to the development of food crops. Traditional plant breeding techniques, such as hybrid production, take many years of development and experimentation. It takes several growing seasons before desired traits can be identified and breeding programmes set up. Gene transfer could allow scientists to select desired characteristics and introduce them directly into crop plants. However, gene identification and transfer in crop plants is still in its infancy and whilst there has been some success with this technique in broad leaved plants such as sugar beet, there has as yet been little success with cereal crops.

9.8.4 *Agrobacterium tumefaciens* and gene transfer in plants

Many broad leaved plants suffer from crown gall disease. This is caused by the bacteria *Agrobacterium tumefaciens* which infects the plant and causes the formation of tumours at the junction of the stem and the shoot. Tumour formation is due to the T_1 plasmid within the bacteria. A fragment of this (TDNA) enters the plant cell and becomes incorporated into the plant's normal DNA. The TDNA will replicate and causes the production of chemicals called **opines** which provide nutrients for the bacteria. The TDNA also causes plant hormones to be produced, leading to the rapid growth of cells and the formation of the tumour. The TDNA fragment is allowed to enter the cell because it recognises a small section of base pairs at each end of the fragment. Scientists have been able to manipulate this naturally occurring process so that desired genes can be introduced into the plant cell. This process is outlined in Figure 93.

9.8.5 Future applications

Much research is currently being performed in the use of gene transfer in plants. Areas of interest include:

● herbicide resistance
● disease resistance
● manipulation of photosynthesis
● manipulation of nitrogen fixation
● improving protein quality.

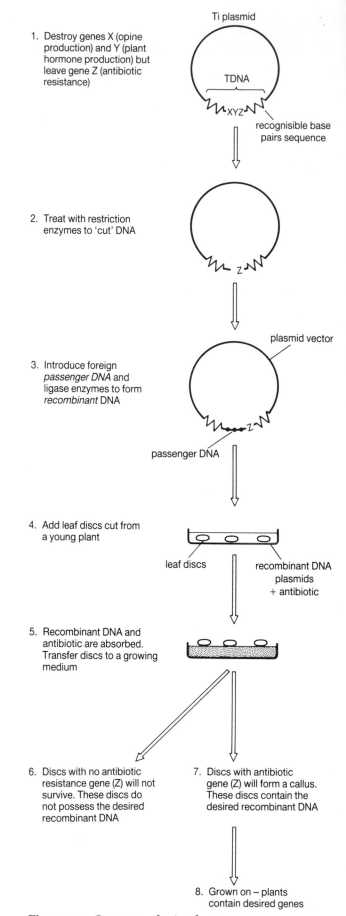

1. Destroy genes X (opine production) and Y (plant hormone production) but leave gene Z (antibiotic resistance)

2. Treat with restriction enzymes to 'cut' DNA

3. Introduce foreign *passenger DNA* and ligase enzymes to form *recombinant* DNA

4. Add leaf discs cut from a young plant

5. Recombinant DNA and antibiotic are absorbed. Transfer discs to a growing medium

6. Discs with no antibiotic resistance gene (Z) will not survive. These discs do not possess the desired recombinant DNA

7. Discs with antibiotic gene (Z) will form a callus. These discs contain the desired recombinant DNA

8. Grown on – plants contain desired genes

Figure 93   Gene transfer in plants

QUESTIONS

1 a) (i) Concisely state what you understand by continuous variation.
   (ii) Give one example of continuous variation.
b) (i) Consisely state what you understand by discontinuous variation.
   (ii) Give one example of discontinuous variaton.
c) List four possible causes of variation.

(Oxford 1991)

2 Primula plants produce flowers which are typically pollinated by insects such as bees. Two different types of flower, pin-eyed and thrum-eyed, are produced. The diagrams show the structure of the two types of flower, and their pollen grains and their stigmatic surfaces.

a) Using the information in the diagram, suggest two ways in which cross-pollination between the two types of flower is favoured.

An experiment was carried out to find out the percentage of successful fertilisation in flowers which were artificially pollinated. The results are shown in the following table.

TYPE OF POLLINATION	POLLINATED FLOWERS WHICH PRODUCED SEED (%)
Thrum pollen on pin stigma	67
Pin pollen on thrum stigma	61
Thrum pollen on thrum stigma	7
Pin pollen on pin stigma	35

b) Describe, briefly, what you would need to do to ensure that pollination occurred only as intended in the experiment.
c) (i) Suggest a non-structural mechanism to account for the lower success of fertilisation between flowers of the same type.
   (ii) Refer to the data in the table. Suggest a hypothesis which might account for the greater success of fertilisation from the artificial self-pollination of pin flower than from thrum flowers.

(JMB 1991)

Structures of thrum- and pin-eyed primula

Vertical section through flower	Pollen grains	Vertical section through the surface of the stigma

Pin-eyed flower

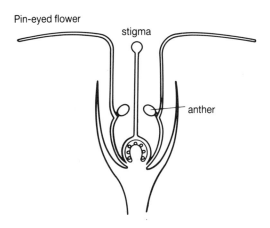

Thrum-eyed flower

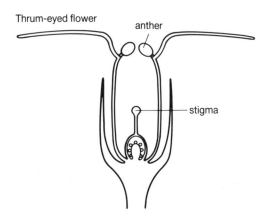

113

9 BREEDING CROP PLANTS

3
a) What is meant by
(i) inbreeding and
(ii) outbreeding?
b) Why does inbreeding increase homozygosity?
c) Why does outbreeding maintain heterozygosity?
d) Imagine you are a plant breeder. Give some examples of situations when
(i) inbreeding is an advantage
(ii) outbreeding is an advantage.

4 The inheritance of two characteristics can be considered at the same time. This is called a **dihybrid cross**. Each characteristic is controlled by a pair of genes which are situated on different chromosomes. They are inherited randomly, in a similar way to those in the monohybrid cross. As seen previously the seeds from *Pisum sativum* can be either round or wrinkled. They can also be yellow or green.

Let R =round
r =wrinkled
Y =yellow
y =green

(yellow is dominant to green)

Each gamete will contain one gene for shape and one for colour. Each plant will contain two genes for shape and two for colour.
a) Assume that pure breeding dominant parent (RRYY) are crossed with pure breeding recessive parents (rryy). List all the possible gametes formed by
(i) the dominant parent
(ii) the recessive parent.
b) What are
(i) the possible genotypes and
(ii) the possible phenotypes of the F_1 generation?
c) Carry out a second cross between individuals of the F_1 generation. List the possible gametes, genotypes and phenotypes of the F_2 generation and the expected ratio of the phenotypes.

5 How might co-dominance increase variation within a population? When might this be an advantage in nature and in the commercial production of animals and plants?

6 Imagine you are a scientist working for a plant breeding institute. You wish to produce seeds which, when sown will give rise to frost resistant cabbages. You could use one of two methods of seed production
(i) pure line selection
(ii) hybrid production.

a) Explain how you would use each method to produce seed.
b) Discuss the advantages and disadvantages of each method.
c) Write a short report for the Chief Scientist explaining which method you would select for producing frost resistant cabbage seed.

7 a) Calculate the variance for each parent and the F_1 and F_2 generations as shown in the example below.

short eared parents

x	frequency, f	x × f	x̄ − x	(x̄ − x)²	(x̄ − x)² × f
5	4	20	1.63	2.66	10.64
6	21	126	0.63	0.40	8.40
7	24	168	− 0.37	0.14	3.36
8	8	64	− 1.37	1.88	15.04
TOTAL	57	x̄ = 6.63			37.44

so N − 1 = 56
variance = 37.44
= 0.67

In this example, we can assume that variation between individuals in the short parent group and in the long parent group is due to the environment only (Ve). This is because these plants are known to be pure breeding and so are genetically the same for this characteristic. This is true of the F_1 plants also as they are obtained from pure breeding parents.

b) Calculate the heritability of ear length in the F_2 generation.
(i) What is the variance calculated for the F_2 generation? This is Vt.
(ii) Calculate the average Ve for this experiment using the figures obtained for the two parents and the F_1 generation.
(iii) You now have figures for Vt and Vt. Calculate variance due to genetics (Vg)

$$Vg = Vt - Ve$$

(iv) Calculate the heritability of ear length using the equation stated.

8 Write an essay describing how gene transfer systems of bacteria and viruses can be exploited by genetic engineers. Your answer should include reference to natural gene transfer processes by bacteria and viruses, ways of inserting foreign DNA into vectors, cloning and the benefits of such genetic engineering.

(JMB 1991)

BIBLIOGRAPHY

Brown, T.A. (1989) *Genctics, A Molecular Approach.* Van Nostrand Reinhold.

Carter, M. (1992) *Genetics and Evolution (Focus on Biology series).* Hodder and Stoughton.

Forbes, J.C., Watson, R.D. (1992) *Plants in Agriculture.* CUP.

Harper, G.H., King, T.J., Roberts, M.B.V. (1987) *Biology Advanced Topics.* Nelson.

Janik, J. (1986) *Horticultural Science.* W.H. Freeman.

Kingsman, S.M., Kingman, A.J. (1988) *Genetic Engineering – An Introduction to Gene Analysis and Exploitation in Eukaryotes.* Blackwell Scientific Publications.

Land, J.B., Land, R.B. (1983) *Food Chains to Biotechnology.* Nelson.

Sofer, W.H. (1991) *Introduction to Genetic Engineering.* Butterworth – Heinemann.

Any standard A level text will deal with Mendelian Genetics.

FEEDING THE WORLD

10.1 FOOD, POVERTY AND POPULATION

Despite the overproduction of food observed in the developed world, the Food and Agriculture Organisation (FAO) of the United Nations (UN) estimates that approximately 15% of the population of developing countries (1.5 billion people) are malnourished or starving. This situation is due in part to the continual growth of the human population.

Since the 18th century the world's population has been growing rapidly and is now increasing by nearly 2% a year. Table 10.1 shows how the world's population has increased since 1930. It is estimated that if this continues the world's population will reach more than 6000 million by the year 2000. This will place increasing pressure on the Earth's capacity to produce food.

Table 10.1   World population size

	WORLD POPULATION (millions)
1900	1600
1950	2500
1975	3967
2045 (*estimated*)	20 000

The food shortage in many countries is reaching dangerous levels. If we are to avoid regular famines, of an increasingly serious nature, then it is essential that efforts are concentrated on ensuring that increases in food production continue at their present rate, or if possible, improve. Agricultural productivity will need to increase at over 3% per year if this demand is going to be met, and agricultural scientists are working to ensure that increases in yield of this size are achieved.

10.1.1 Nations at risk

The nations at greatest risk of severe problems due to hunger and poverty have several characteristics in common:
* many nations are small. Of the 161 nations at greatest risk, 105 have populations of less than 10 million (UK population is approximately 50 million).
* many nations are newly independent. Of the 43 nations most seriously affected by poverty and hunger, 36 have become independent since 1945, 29 of these since 1960.
* nations are extremely poor. Of the nations at greatest risk, 61 have a Gross National Product (GNP) of $500 per capita, per year, compared with a GNP of over $7,000 in the USA. To add to this problem the incomes of people living in rural areas tend to be well below the national average, creating pockets of increased poverty.

- individual farm sizes are tiny, compared to those in developed areas. For example, 92% of all farms in Nepal are less than three hectares in size.

Small nations face many of the problems faced by the larger nations of the developing world. However, because of their size, they are also unable to provide a full range of scientific technical and financial services within the country and so must rely on help from elsewhere.

Food production in newly independent nations has often been managed by the colonial powers which once controlled them. These nations are often producers of cash crops such as coffee, cocoa and rubber for export. The growth of food crops has traditionally been of low priority.

In many developing regions there is the added problems of civil war and military rule which hinders food production.

10.1.2 Climatic problems

Figure 94 shows the climatic regions of the world. Most developing regions lie with the tropics.

The tropics lie between the tropics of Cancer and Capricorn, 23.4°N and 234°S of the equator. In general, temperatures rarely fall below 18°C. The average temperatures of this region are shown in Table 10.2.

Figure 94   Climatic regions of the world

Table 10.2   Mean temperatures in the tropics

LATITUDE (degrees)	MEAN TEMPERATURE (°C)
20–30 N	20.6
10–20 N	25.3
0–10 N	25.7
0–10 S	25.0
10–20 S	23.5
20–30 S	19.0

(Sellers 1965)

The soil temperature is also high, ranging from 26–35 °C. This increases the rate of processes such as mineralisation and evapotranspiration (see Chapter 5).

Rainfall in the tropics tends to be heavy and seasonal. The amount and duration of the rainfall tends to determine which crops are grown. At the equator, rain falls almost every day, but as you move away from this region rainfall patterns vary. In many areas rain only falls during one time of the year. This rainy season may last from 50 to more than 200 days.

These climatic features mean that crops grown in these areas must be suited to high temperatures and be able to withstand periods of both heavy rainfall and drought.

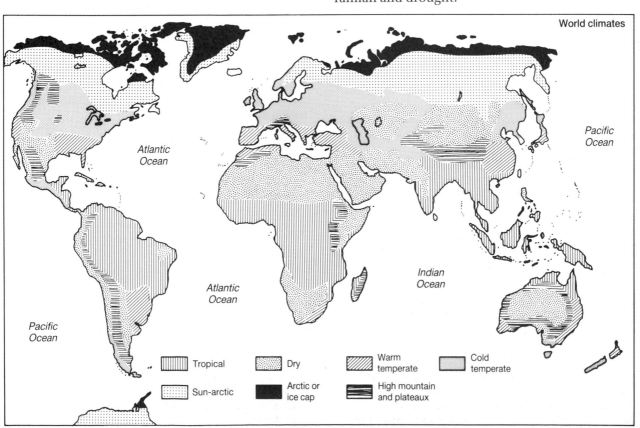

World climates

▯ Tropical	▯ Dry	▯ Warm temperate	▯ Cold temperate
▯ Sun-arctic	■ Arctic or ice cap	▯ High mountain and plateaux	

10.2 INCREASING FOOD PRODUCTION IN THE DEVELOPING NATIONS

There are two main lines of investigation in trying to improve food productivity in developing nations:

(i) We can increase the area of land from which food is produced.

(ii) We can increase the productivity of existing areas of land.

10.2.1 Increasing areas of cultivation

Worldwide, about 1.3 billion hectares of land are used for crop production. However, there are still large areas of land which could be used for growing crops, which are left uncultivated. This is illustrated in Table 10.3.

Table 10.3   Present and possible arable land

REGION	CULTIVATED LAND ($\times 10^6$Ha)	POSSIBLE ARABLE LAND WHICH IS UNCULTIVATED ($\times 10^6$Ha)
Africa	160	600
Asia	500	300
Americas	300	500
Australia	20	500
Total	980	1900

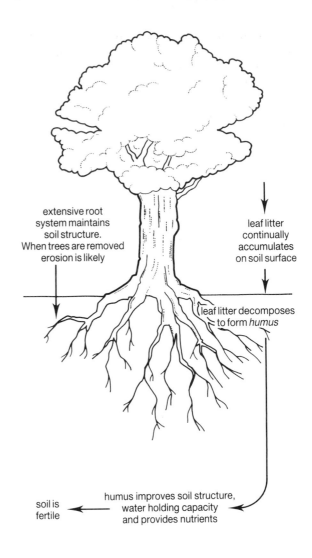

extensive root system maintains soil structure. When trees are removed erosion is likely

leaf litter continually accumulates on soil surface

leaf litter decomposes to form *humus*

soil is fertile

humus improves soil structure, water holding capacity and provides nutrients

Figure 95   Soil structure and fertility is maintained within the rain forest

Much of the uncultivated land is in the tropics where environmental factors such as rainfall patterns may prohibit or limit crop growth. In some cases the cost of clearing land and building roads so that crops can be transported to market is so high that cultivation is not economic.

Large areas of the tropics are covered in diverse rain forest. Leaf litter from this vegetation falls onto the soil surface and decomposes rapidly due to the higher temperatures, replacing the nutrients used by the plants for growth. The high content of organic matter also helps maintain the soil structure, as does the presence of large tree roots which help prevent erosion.

When areas of tropical rain forest are cleared for agriculture, the trees are cut down and burned to expose bare soil. Tropical soils tend to weather rapidly due to the higher temperature and humidity. This means that soils have a deep A horizon (see Chapter 5) which lets water percolate through rapidly. Although tropical soils are – initially – suitable for crop production, continual heavy rain leaches nutrients from the topsoil making them unavailable for plant growth. During the dry season, the soil dries out and becomes very hard. When heavy rain falls on this dry soil, there is a danger of erosion and soil loss.

Newly cleared tropical soil is extremely fertile due to

- the organic matter contained within the soil due to the decomposition of leaf litter,
- the nutrients added to the soil as a result of the burning of trees.

However, after a few years the fertility level of the soil declines because nutrients are being removed by plants for growth and are also subject to leaching. Leaching makes it difficult to maintain the fertility level of the soil. The most successful method involves treatment with inorganic fertilisers and organic fertilisers in the form of grass mulch. (See Table 10.4.)

Table 10.4   The yield (kg/Ha) of cotton on cleared tropical rainforest

YEAR	NO ORGANIC FERTILISER OR MULCH	INORGANIC FERTILISER ONLY	MULCH ONLY	BOTH
1	200	398	120	1430
2	180	790	1460	1980
3	120	700	980	1340

10.2.2 Increasing productivity of existing land

Since the Second World War, the developing nations of the world have strived to increase food production through the use of agrochemicals, new technology and genetically improved crop plants. This process has been called the **Green Revolution** and has been so successful in developed regions that food production is now controlled by legislation to prevent overproduction and so maintain stable prices (see Chapter 7). The following section examines the impact of the Green Revolution on developing regions.

10.3 THE GREEN REVOLUTION IN DEVELOPING REGIONS

The Green Revolution occurred in the 1950s and 1960s in an attempt to rapidly increase the amount of food available to feed a growing population. It had two main aims:
- to set up breeding programmes to produce early maturing, high-yielding varieties of crops which could withstand the climatic conditions experienced in the tropics.
- to introduce technologies such as mechanisation, the use of agrochemicals and irrigation into developing regions.

The implementation of these aims has led to an increase in food production by 7% in developing countries since the mid-1960s (see figure 96) but this gain has not been without its problems.

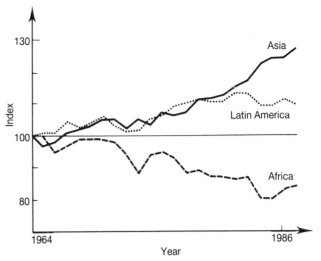

Figure 96   Changes in per capita food production in the developing world

10.3.1 Crop improvement

As we saw in Chapter 9, a knowledge of the mechanisms of hereditary allows scientists to select some of the characteristics which an individual exhibits. The aim of crop improvement is to produce new strains or varieties which have high yields, a better nutritional value or are more resistant to pests and diseases than existing ones.

Figure 97 illustrates some of the characteristics which may be included in a breeding programme.

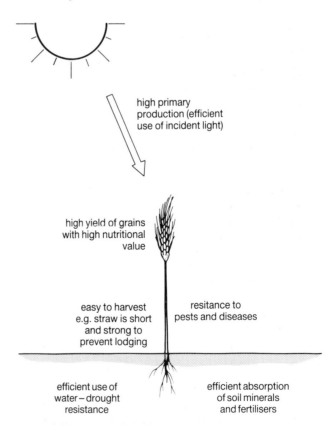

Figure 97   Some possible aims of crop improvement in wheat

PLANT SCIENCE IN ACTION

10.3.2 High-yielding varieties

Since the 1950s, new high-yielding varieties of both rice and wheat have been developed. Between a third and a half of the rice areas in developing nations are now planted with high-yielding varieties of rice. The introduction of these varieties has increased crop productivity. However many farmers have substituted these crops, grown in monoculture, for the more traditional mixed crops which have not been so highly developed. This leaves farmers more prone to the problems associated with monocropping and rural communities relying on the sale of these crops to generate income to buy other foods.

10.3.3 High-yielding wheat

Traditionally, wheat was bred to produce plants with tall stalks. This made the grain easy to harvest by hand. However, tall stalks tend to be more susceptible to wind damage, causing them to lodge or fall down. This makes mechanised harvesting difficult. The main objectives of wheat breeding programmes were therefore:
- to produce high yielding strains
- to produce strains with short stems.

Long stemmed wheat varieties are susceptible to lodging.

The development of modern wheat strains began in Japan. In 1917 the Japanese crossed the native Daruma wheat, which was a dwarf variety, with a high-yielding American strain called Fultz. Short-stemmed, high-yielding offspring were crossed again with another high-yielding American wheat called Turkey Red. By the mid 1930s, a high-yielding, short-stemmed wheat variety called Norin 10 was available.

After the Second World War, Norin 10 was taken to America where scientists carried out further crosses to produce a short-stemmed variety suited to the local environmental conditions called Gaines. Gaines was, however, unsuitable for growth in tropical regions as it was susceptible to rust infections. Rusts are fungal diseases which thrive in the hot and humid tropical climate. Gaines was also a winter wheat which required vernalisation to stimulate grain production.

In 1953, an American, Norman Borlang, and his team started work in Mexico on the development of high-yielding varieties of wheat which would be suitable for growth in the tropics. Ten years later they had developed high-yielding, short-stemmed strains which were resistant to rust infections. These varieties were spring wheats which did not require vernalisation and matured quickly so it was possible to grow two crops a year. In 1963, 100 kg of seed was sent to the Indian government and now more than one third of India's wheat and one half of Pakistan's is derived from these varieties. Norman Borland won the Nobel Peace Price for this work in 1970.

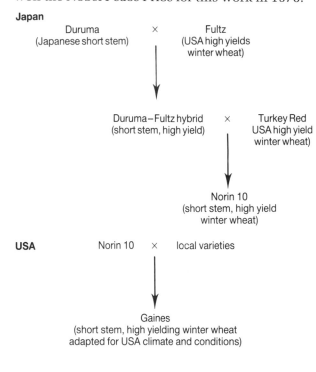
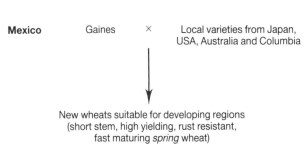

Figure 98   The development of high yielding varieties of wheat

120

10.3.4 High-yielding rice

Tradition rice varieties are well adapted to tropical climates and rainfall patterns, and to low soil fertility. However, like traditional wheat, the plants are tall and slow to mature so only one crop can be grown each year and yields are affected by lodging.

Breeding programmes in several areas of the world led to the development of the rice strain IR–8 by the International Rice Research Institute in the Philippines. This rice variety had a short stem, so it did not lodge and was high-yielding, however the grains remained hard after cooking. By 1971 a softer grained variety IR–24 was produced. Further developments have produced rice varieties which are high-yielding, respond well to fertilisers, are disease resistant and mature quickly.

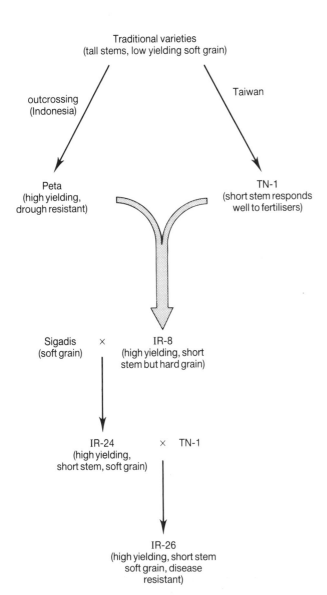

Figure 99   The development of high yielding varieties of rice

10.3.5 New technologies

The improved yields obtained by the use of high-yielding varieties of crops are only possible if they are accompanied by other technological inputs, such as irrigation, mechanisation and the use of agrochemicals.

10.3.6 Irrigation

In many areas the growth of high-yielding varieties of wheat and rice are limited by the supply of water. Both wheat and rice require a large amount of water at certain stages of their life cycle and so the introduction of these crops is limited to certain regions only.

Irrigation allows the water supplied to a crop to be controlled. Irrigation is effective but expensive, especially when water is scarce. Table 10.5 compares the yields of rice obtained from high yielding varieties under different conditions in the Philippines.

Table 10.5   *Yields of irrigated rice*

CONDITIONS	YIELD (t/Ha)
Fully irrigated	2.5
Part irrigated	1.9
Rainfed	1.6

Many traditional irrigation systems have been extended and new ones developed using machinery to bore wells and lay pipes. However many tropical areas are unsuitable for irrigation. This is because they have soil which is susceptible to erosion or leaching or simply because there is no natural water supply available during the dry season. In an attempt to overcome these problems, scientists are trying to develop drought resistant strains of crop plants.

10.3.7 Fertilisers

High-yielding varieties of wheat and rice produce higher yield than normal varieties without the use of fertilisers. However, if fertilisers are applied to crops, the response in terms of yield is highly significant. This is illustrated in Figure 100.

The availability and use of inorganic fertilisers has been a major factor in the increase in food production. The Food and Agriculture Organisation (FAO) estimates of world fertiliser consumption are shown in Table 10.6.

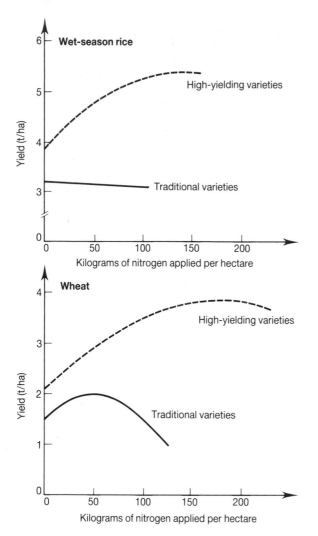

Figure 100   The response of wheat and rice to fertiliser

Table 10.6   World fertiliser consumption (excluding China)

YEAR	FERTILISER CONSUMPTION $\times 10^6$t
1900	2
1914	4
1938	9
1945	7
1955	21
1965	41
1975	90

Inorganic fertilisers are expensive to produce and therefore, expensive to buy. This means that small farmers in developing nations are often unable to buy and use inorganic fertilisers unless they are producing high value cash crops such as tobacco or coffee for export.

10.3.8 Pesticides

The demand for pesticides has risen by 20% a year. Most pesticides are relatively expensive and so, like fertilisers, are limited to use on high value crash crops. However, the use of pesticides in high-yielding varieties can produce a great increase in yield and, as a result, profit.

Table 10.7 shows how technological advances can enhance the yields obtained from high-yielding varieties of rice in Asia.

Table 10.7   The enhancement of yields obtained from high-yielding varieties of rice in Asia by the use of modern technology

INNOVATION	INCREASE IN TOTAL RICE YIELD PER YEAR
High-yielding varieties alone	30 million tonnes
Fertiliser application	30 million tonnes
Irrigation application	35 million tonnes

10.4 BEYOND THE GREEN REVOLUTION

The implementation of the aims of the Green Revolution has led to a 7% increase in food production by developing countries since the mid 1960s. This gain has not been without its problems.

10.4.1 Changes in cropping systems

The development of high-yielding varieties of rice and wheat has been accompanied by a change from traditional cropping systems to monocropping. Traditionally farmers in developing nations grew a mixture of crops to support their family or village by intercropping. This allowed them to produce a variety of foods and ensure against starvation if one crop was to fail.

Farmers have been encouraged by the high yields obtained from the new varieties of wheat and rice and so have tended to substitute these cereals for the crops they traditionally grew.

Monocrops are more suceptible to attack by pests and so there is an increased requirement for pesticides. The use of pesticides sometimes makes matters worse as with the case of the brown planthopper infestation which occurred in rice in the 1970s. The brown planthopper is naturally controlled by the wolf spider – an example of biological control. However, the use of pesticides wiped out the wolf spider in some regions, leaving the brown planthopper to destroy 2×10^6 t of rice in Indonesia in 1977.

10.4.2 Crop choice

In nature, plants are only found in habitats which suit their needs in terms of soil type and climate. In agriculture, plants are sometimes sown in regions which are not suitable.

Table 10.8 shows the climate suitability of the six main cereal crops grown worldwide. Figure 101 shows the distribution of these crops.

Table 10.8 Climate suitability of cereals

CROP	TEMPERATURE (°C)	RESPONSE TO DROUGHT	METHOD OF PHOTOSYNTHESIS
Wheat	<20	susceptible	C_3
Barley	<20	susceptible	C_3
Rice	>20	susceptible	C_4
Maize	>15	susceptible	C_4
Sorghum	>20	susceptible	C_4
Millet	>20	resistant	C_4

10.4.3 What makes wheat unsuitable?

It can be seen that wheat is quite commonly grown in tropical areas, especially in Asia. Farmers have been encouraged to grow wheat because it produced high quality flour and can command a high market price. However, most wheat varieties are not suited to the tropical environment for a number of reasons.

- Wheat is a C_3 plant and so photorespiration will occur rapidly at high light intensities (see Chapter 3).
- Wheat is susceptible to drought – in tropical regions it requires irrigation if high yields are to be obtained (4–5 cm of water biweekly).
- Wheat grows best at temperatures below 20 °C. It cannot tolerate temperatures above 29 °C. Varieties of wheat with short growing seasons are, therefore most successful as they can complete their life cycle before it becomes too hot.

A combination of these factors means that the growing of wheat is often confined to areas of high altitude where rainfall is higher and the temperature lower. Yields often do not reach their full potential: the average yield of wheat in the tropics is 1.35 t/Ha compared with about 7.5 t/Ha in temperate climates.

Figure 101   The distribution of cereal crops

10.4.4 The suitability of native crops

Tropical areas are more suited to the growth of native crops such as sorghum and millet. Both sorghum and millet are cereals which grow well in arid and semi-arid areas. They provide the staple food of 90% of the rural population of Africa. Both cereals are used to produce flour, but the flour is of a lower quality than wheat flour. Neither sorghum or millet are very marketable, except as animal feed.

Sorghum

Millet

Millet is especially well adapted for tropical environments. It is a C_4 plant which grows best at temperatures above 20 °C. Millet can grow on very infertile soils, although it responds well to the use of fertilisers. Millet's best quality is its drought resistance. Millet has an extremely extensive root system which allows it to extract water from deplected soils efficiently. It also has few stomata which, it is believed, close quickly when water is short slowing down transpirational losses. Millet uses water efficiently, producing 3.6 g of dry matter for each kilogram of water it absorbs. This compares with a figure of 1.4 g in rice. Drought resistance is also aided by its cell structure. In millet, the cellulose cell wall is impregnated with silica which it obtains from the soil. This strengthens the cell walls so they are less likely to become fully plasmolysised, even under conditions of extreme water stress.

However, millet has not been subjected to the intense development programmes set up to improve wheat varieties. It is still a very low-yielding crop and the flour it produces is of poor quality, lacking several essential amino acids. Although it is suited to the tropical environment, these factors, alongside its low market value, discourage farmers from growing millet in favour of non-native crop plants.

10.4.6 Appropriate technology

New varieties of cereal crop can only produce their potential yield if they are supported by the necessary conditions. These include:
- water – rainfall is often insufficient and so irrigation is required,
- nutrients – soils are often lacking in nutrients due to continual cultivation or leaching and inorganic or organic fertilisers are required,
- protection from pests – the use of chemical pesticides and other control methods,
- mechanical planting, cultivation and harvesting – the application of water and chemicals requires specialised machinery. New crop varieties have been developed so that they are easy to harvest mechanically.

The manufacture and running of these technological inputs requires large amounts of energy. The availability and cost of fuel which is required to run machinery, such as tractors and irrigation pumps, often prohibits their use on small farms. The owners or tenants of small farms cannot afford to buy all the components required to produce maximum yield from the new varieties of crop. On most small farms human labour still replaces the use of machinery and good crop husbandry, for example, careful choice of cropping system reduces the need for fertiliser and pesticides. In many cases, small farmers who employ these traditional methods can obtain yield comparable with, or even higher than, those obtained by larger more intensive farms. The reason for this is unclear. However it is thought that small farmers take great care over their crops as they are often the main supply of food for their family. The use of low levels of technological inputs, such as fertilisers, to support traditional techniques can increase yields without the farmer having to invest in more expensive technology.

10.4.7 The Green Revolution and the small farmers

Traditionally, the majority of farms in developing countries are small. For example, in the Punjab region of Pakistan, 79% of farmers cultivate only 32% of the land in holdings of less than four hectares. Most small farmers do not own the land that they farm but are tenants. Tenant farmers rent their land in one of two ways:

(i) Sharecropping – the farmer works on the landowner's land in return for a fixed proportion of the crops grown.

(ii) Renting – a fixed rent is paid to the landowner.

The Green Revolution has affected the small farmer in several ways. Many small farmers have been unable to buy the technological innovations required to compete successfully with the larger farms. They have been forced out of business and large numbers of people have migrated to the cities in search of work. This adds to the problems of high unemployment, overcrowding and poverty already experienced in many cities.

The Green Revolution has also upset the relationship between landowner and sharecropper. Landowners who have been able to increase their yields by using new technologies are unwilling to continue giving a fixed proportion of yields to the sharecroppers as the increases in yield mean that they can now sell their surpluses on the open market. Sharecroppers do not benefit from this arrangement as they are only supplied with enough of the crop to provide food for their families.

10.5 THE WAYS FORWARD?

If we are to avoid regular famines in the developing world, there must be a move towards the **sustainability** of agriculture in these regions. Agricultural sustainability can be defined as the ability to maintain productivity, despite environmental and climatic factors. Unless the strategies introduced to increase productivity are environmentally and socially sustainable, lasting improvement cannot occur. Sustainability requires national policies, incentives, development projects which may be externally funded but which are consistent with the constraints of the local environment, local traditions and values. Some important approaches are summarised here.

- The reduction of the rate at which the human population is expanding by means of national implementation of family planning and other health programmes.
- Research into the improvement of native crops such as sorghum and millet.
- The supply of aid from developed nations in the form of investments which encourage long term projects.
- Greater economic co-operation between developed and developing nations which helps reduce the burden of debt experienced by many Third World countries and removes trade barriers so that developing nations are able to trade freely with the West.

QUESTIONS

1 The table below shows the increase in human population size since the year 1 AD.

Increase in world population

DATE	ESTIMATED POPULATION (millions)
1	250
1650	500
1850	1 000
1930	2 000
1976	4 000

(From: *Plants, Food and People* by Chrispeels and Sadava. Copyright © 1977 by W. H. Freeman and Company. Reprinted with permission.)

a) Plot these figures on a graph.

b) The data shows the number of years it has taken for the population to double since the year 1 AD. What trend do you notice?

c) What do you think is the significance of this trend?

2 a) The soil found in a tropical rain forest is very fertile. Explain how the structure and fertility of this soil is maintained.

b) The figure on page 126 show a profile of a section of soil from a tropical rain forest.

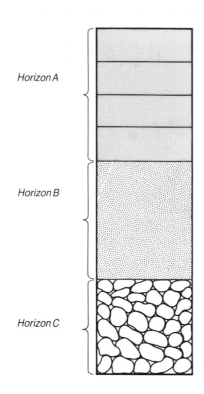

Horizon A

Horizon B

Horizon C

Soil profile

     (i) Identify the layers A, B and C.
     (ii) Humus is produced in layer A. Explain how.
     (iii) Why does soil from tropical rain forests have a high humus content?
  c) When tropical rain forest is cleared for agriculture why is soil fertility
     (i) improved initially?
     (ii) decreased in subsequent years?
  d) Imagine you were asked to talk to subsistence farmers in the tropics. Explain the implications of the slash and burn techniques and suggest methods of improving tropical soils for long term cultivation.

3 The EC controls food production in Europe through the Common Agriculture Policy. This prevents overproduction by paying farmers subsidies if they limit the amount of food they produce (Chapter 7). Discuss the ethics of this policy in the light of the information contained within this chapter.

4 'The Green Revolution has not fulfilled its potential within the developing world' – discuss.

5 Write a report to the United Nations suggesting ways for Third World agriculture to move forwards after the Green Revolution.

BIBLIOGRAPHY

Borgstrom, G. (1965) *The Hungry Planet.* Macmillan.

Brown. L.R., Finsterbusch, G. (1972) *Man and His Environment: Food.* Haper and Row.

Chrispeels, M.J., Sadava, D. (1977) *Plants, Food and People.* W.H. Freeman.

Conway, G.R., Barbier, E.B. (1990) *After the Green Revolution.* Earthscan Publication Ltd.

Forbes, J.C., Watson, R.D. (1992) *Plants in Agriculture.* CUP.

Jurion, F., Henry, J. (1969) *Can Primitive Farming be Modernised?* Congo Institute of Agricultural Studies.

Pinstrup-Andersen, P., Hazell, P.B.R. *The Impact of the Green Revolution and Prospects for the Future.*

11 ALTERNATIVE FOOD SOURCES

11.1 TRADITIONAL FOOD SOURCES

Most of the food that we eat is produced by traditional agricultural methods. Only a small number of plants and animals are used for food. However, an ever increasing human population means that the world's traditional food sources are being put under continual strain and other sources must be developed. Two thirds of the world's population has a deficient diet. The most severe deficiency is with respect to lack of protein, or lack of high quality protein.

11.2 PLANT PROTEINS – THE ADVANTAGES AND PROBLEMS

Plant protein has several advantages over other alternative protein sources. Plants are cheap to produce and they are an acceptable food source for many ethnic and religious groups. But the protein within a plant is often difficult to extract due to the presence of the cellulose cell wall. Plant protein tends to be of a lower quality than animal protein, as it is deficient in some of the essential amino acids. A further problem is that many plant proteins are associated with toxins. These toxins may cause dramatic effects as the examples below show:

(i) The Lathyrus pea is a drought resistant legume which forms an important part of the diet of people living in very dry areas. It contains a neuro toxin which accumulates in the body. If more than 300 g of the peas are eaten daily the motor neurones become damaged and the lower body will become paralysed.

(ii) Peanuts are associated with the growth of a fungus *Aspergillus flavus*. This fungus attacks the nuts inside pods which have been broken or damaged and produces a toxin called alfatoxin. This is thought to be carcinogenic, although there have been no ill effects reported in humans.

Legumes, for example peas, lentils, peanuts and kidney beans, have been part of the human diet

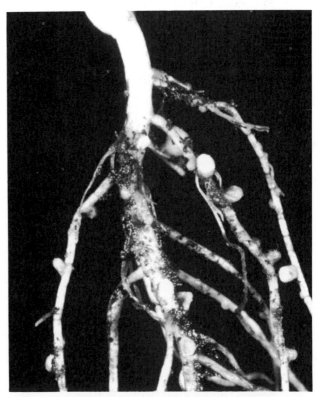

Root nodules of legumes contain symbiotic bacteria.

for thousands of years. They are a useful protein source because they are able to produce large amounts of edible protein without requiring nitrogenous fertilisers. This is because of the symbiotic relationship which exists between the plant and micro-organisms which live in the nodules found on their roots. The micro-organisms extract carbohydrate for their metabolism from the plant and in return, they produce nitrates which are released into the soil for uptake by the plant.

There is no need to increase the production of legumes in developed countries. However many of the developing nations depend on starchy cereal crops such as sorghum or rice to provide their staple diet. In these areas the increased production of legumes would have a considerable impact upon the diet.

Table 11.1   Yield of protein from different crops

CROP	PROTEIN (kg/Ha)
Rice (starchy)	42
Sorghum (starchy)	33
Maize (starchy)	73
Sweet potato (starchy)	37
Chick peas (legume)	90
Groundnut (legume)	113

11.2.1 Soyabeans

Soyabean is an excellent source of protein. Its amino acid content compares well with that of egg (except with respect to the amino acid methionine). Soyabeans are used extensively in human foods but they must be processed carefully before they are suitable. Boiling soyabeans leaves them hard and bitter, and likely to cause diarrhoea due to the presence of the trypsin inhibitor.

Soyabeans can be soaked and broken up so that an emulsion forms. If this is heated soya curd is produced. This contains 8% protein, 3% fat and 1.5% carbohydrate. This form of soyabean is eaten extensively China and Japan.

Soyabeans can be fermented to produce a product called temph. This is commonly produced in Japan. Cooked soyabeans are innoculated with a fungus *Rhizopus oryzae* and left to ferment for 18–24 hours. Fermentation lowers the amino acid content slightly but enhances the palatibility of the soyabeans.

Soyabeans can be used to produce meat analogues. Meat analogues are cheap to produce

and have a longer shelf life than meat. Soyabeans are milled to produce soya flour. From this protein is extracted by treatment with food grade alkalis. The protein is then forced through a membrane impregnated with tiny holes. As the protein jets emerge (through the holes) they are coagulated in an acid bath so that thin fibres are formed. These fibres are stretched and coated with flavouring, colouring and other nutrients if required. They are then bound together to form meat analogues. This process is shown in Figure 102. The resulting product usually contains between 30–69% protein.

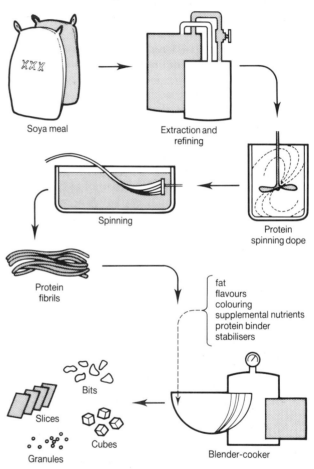

Figure 102   Producing soyabean analogues

11.2.2 Groundnuts

Groundnuts (peanuts) are extensively grown in tropical and semi-tropical regions. They were originally introduced from South America for growth as an oil-producing crop. Groundnut kernels contain 25–30% protein. This is mainly made up of the amino acids aspartic acid, glutamic acid and arginine. Groundnuts, like soyabeans, are also deficient in methionine.

Roasting can profoundly alter the nutritional value of ground nuts. If the nuts are roasted slowly the content of some amio acids is diminished. However, if they are roasted more quickly using high temperatures, the amino acid content is less affected although amino acid availability may be lower.

Half of the groundnuts produced annually are used to produce peanut butter. This has a protein content of 25% and contains about 6000 calories a kilogram!

11.2.3 Leaf protein

Leaves contain only a small amount of protein which is largely unavailable to humans because we are unable to digest the cellulose cell walls which bind plant cells. It is possible however, to extract the protein from leaves and produce a concentrate of leaf protein which is suitable as both animal feed and for human consumption.

Many types of leaf can be processed in this way and so the raw material can easily be obtained either by cultivation or from the wild. The process is relatively simple:

- the leaves are collected and cleaned
- they are ground to a pulp
- they are pressed to extract 'juice'
- the juice is boiled quickly to coagulate the protein
- the curd is pressed to remove water and leaves the green crumbly leaf-protein concentrate.

The production of leaf-protein concentrate
a: Feeding the cleaned leaves into the pulper.
b: Grinding the leaves into a pulp.
c: Extracting the juice.

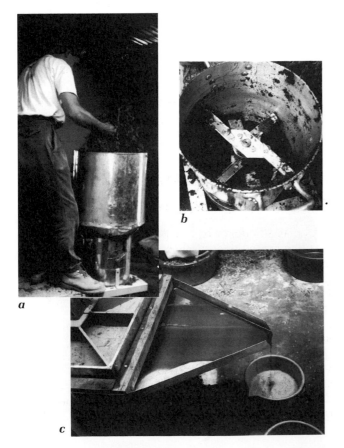

Small processing plants have been set up in villages throughout the Third World enabling people to produce leaf-protein concentrate for use in their own community. Feeding trials on children show the dramatic effects of adding leaf-protein concentrate to the diet.

Children drinking leaf protein concentrate

In many areas where leaf-protein concentrate is produced, villagers will grow one leafy crop for protein production and one traditional carbohydrate crop such as cassava. This allows a balance of protein and carbohydrate to be obtained.

Now try Investigation 26 Making Leaf Protein Concentrate in the *Plant Science in Action Investigation Pack.*

11.3 SINGLE-CELLED PROTEIN

Single-cell protein is protein derived from the whole cells of micro-organisms. Microbes have a high protein content (40–50% dry matter), and with their rapid rate of reproduction, they have the potential to produce large quantities of protein. Microbes require a source of energy or a substrate so they are able to reproduce and produce protein. A wide range of waste products from the petrochemical and food industries produce suitable substrates for many microbes.

Most sources of single-cell protein, including those mentioned in the following section, are more suitable as animal feeds than as food for humans. They have a high content of nucleic acid which humans are unable to digest in large quantities. There are also problems associated with the marketing of such products as people are often unwilling to eat products made from 'germs'!

11.3.1 Algae

Algae are single-celled organisms which are able to photosynthesise. This makes them an extremely useful source of protein as they use solar energy to produce chemical energy for growth. Under production conditions it is possible to produce 30 times as much protein from algae than it is from the equivalent area of a field of beans.

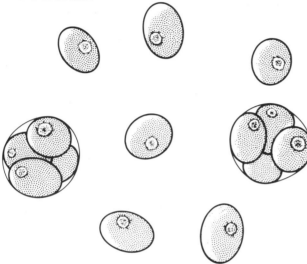

Algae can be used to produce protein

The most commonly used algae for protein production is *Chlorella*. *Chlorella* can be grown in culture medium containing a mixture of mineral salts. It is illuminated so that the cells will photosynthesise. The maximum yield is about 34 g a day from a initial culture of 10 g of *Chlorella*. The main problem associated with the production of *Chlorella* is that the cells will only grow in thin layers to ensure constant illumination.

Chlorella cells are harvested, washed and dried to produce a green powder which tastes like green tea. It has a protein content of about 42%.

11.3.2 Yeast

Brewer's yeast contains 40–50% protein. Most of this is in the form of enzymes. The yeast *Saccharomyces cerevisiae* contains all the essential amino acids and many other nutrients, such as the B vitamins, and some minerals.

Yeast requires a source of carbohydrate for growth. This can be supplied from many inexpensive waste products such as molasses and wastes from the fruit, forestry and dairy industries.

During the war, the yeast *Torula utilis* was used to supplement certain foods. The yeast was grown for several hours, collected and washed and then dried to produce flakes which had a nutty flavour. This could be added to foods such as stews and soups.

Since the 1960s British Petroleum has been producing high grade yeast protein from petroleum. Their production plant is capable of producing 20 000 t of protein a year. The yeast (*Candida lipolytica*) uses the wax fraction of crude oil (a by-product of the petrol industry), as its substrate. This process is highly exothermic and so the substrate needs to be continually cooled so that the optimum temperature for the yeast is maintained.

11.3.3 Bacteria

Bacteria contain between 40–80% protein (dry weight). The content of essential amino acids will vary with species. A system has been developed for producing dry cells of *Bacillus megaterium* on molasses mainly for the production of B vitamins. However, in the 1970s, ICI successfully manufactured a protein from bacteria. The bacterium *Methylophilus methylotrophus* was developed as a source of single-cell protein. *M. methyloptrophus* was grown on methanol and produced a product called Pruteen.

11.3.4 Fungi

The most common type of fungi used for food is the mushroom. Mushrooms are a good source of protein but are deficient in the amino acid tryptophan.

Another type of single-cell protein which has been successfully used as a human food is derived from the fungi *Fusarium graminearum*. It is called **myco-protein**. Myco-protein is treated, during production to reduce the nucleic acid content and, after further processing, the end product 'Quorn®' compares well with meat – being only slightly lower in protein, higher in fibre and lower in fat.

Myco-protein

QUESTIONS

1 a)   (i) What is myco-protein?
     (ii) State the name of one organism used to provide myco-protein.
     (iii) Describe a medium in which this organism might be grown.

  b) List three advantages of producing myco-protein compared with conventional methods of producing protein.

  c) State two disadvantages of using protein produced by micro-organisms for animal foodstuffs.

  d)   (i) Fungi are used to produce synthetic meats for human consumption. What are the structural features of fungi which have made them suitable for this purpose?
     (ii) Suggest two reasons for the public reluctance to accept in their diet protein food products made by micro-organisms.

(UCLES 1992)

2 Companies producing single-cell protein have had problems with marketing it.
  a) Why do you think this is so?
  b) What are the features of single cell protein which could be used as a basis of a good advertising campaign?

3 Leaf-protein concentrate can be produced from almost any leaf. What are the added advantages of growing a leguminous crop such as alfalfa or beans as a source of leaf?

4 The table below shows the percentage of the recommended daily intake of nutrients which can be supplied to a child by feeding 40 g of leaf-protein concentrate a day.

  a) Half a billion children, and 60% of all pregnant women in the developing world are anaemic. How would eating leaf-protein concentrate improve the health of these people?

  b) Leaf-protein concentrate contains 27% of the RDA of protein. How might the remaining requirement be obtained?

  c) Leaf-protein concentrate is commonly used to supplement the diet of children. Why is the treatment of children targeted?

NUTRIENT	PERCENTAGE OF RDA
Vitamin A	460
Iron	100
Vitamin E	61
Calcium	37
Niacin	30
Protein	27
Magnesium	51
Folic acid	70
Copper	29
Zinc	14
Riboflavin	8
Thiamin	8

5 The table below shows the amino acid content of egg and soyabeans.

AMINO ACID	ISOLATED SOYABEAN PROTEIN g/16g N	WHOLE EGG g/16g N
Arginine	8.3	6.6
Cysteine	0.7	2.3
Histidine	2.6	2.4
Isoleucine	6.5	6.8
Leucine	7.5	9.0
Lysine	6.8	6.3
Methionine	1.0	3.1
Phenylalanine	5.0	5.9
Threonine	3.9	5.0
Tryptophan	1.0	1.7
Tyrosine	3.4	4.4
Valine	5.5	7.4

a) Draw a bar chart to show each of these data. Whole egg is a good source of protein. Compare soyabean protein to egg protein and comment on its potential as a protein source.

b) Imagine you are working for VSO (Volunteer Services Overseas) in a village in Africa. You decide to encourage the farmers to grow soyabeans to supplement their traditional crops of sorghum and millet.
   (i)   What are the advantages of this move?
   (ii) What problems might the farmers encounter?
   (iii) Suggest a suitable method of processing the soyabeans which can be used within the village to produce a palatable, digestible food.

6 Groundnuts were originally planted in many regions so that they could be harvested for their oil. It is mainly the countries which suffer from protein malnutrition which grow ground nuts for oil extraction and export. Why do you think groundnuts are grown for oil (as a cash crop) and not as a source of protein for the people living in these regions?

BIBLIOGRAPHY

Chrispeels, M.J., Sadava, D. (1977) *Plants, Food and People.* W.H. Freeman.

Freeland, P. (1992) *Micro-organisms in Action (Focus on Biology series).* Hodder and Stoughton.

Martin, C. *All Grass in Flesh.*

A selection of leaflets, books and videos about leaf-production concentrate can be obtained from:

   Leaf For Life,
   Find Your Feet,
   37–39 Great Guildford St,
   London, SE1 0ES.

GLOSSARY

Abscissic acid: A plant growth hormone particularly important in seed dormancy.

Abscission: The dropping of leaves.

Absolute Growth Rate: The amount by which a plant has increased in size during a given period of time.

Agrochemicals: General name given to fertilisers, pesticides and other chemicals used in agriculture and horticulture.

Aleurone layer: The outer layer of the endosperm of a cereal grain.

Algae: Simple plants that usually live in water. A source of single-cell protein.

Annual: A plant which completes its life cycle in a year.

Asexual propagation: Propagation which does not depend on the joining of male and female gametes.

Autotroph: An organism which can produce its own food from inorganic molecules. Usually a plant.

Auxins: A group of plant growth regulators involved in most aspects of plant growth. Synthetic auxins are used commercially to manipulate growth in plants.

Available water: The amount of water in soil which can be absorbed by the plants growing there. Calculated by subtracting the permanent wilting percentage from the field capacity.

Biennial: A plant which completes its life cycle in two years.

Biological control: The use of natural methods in pest control e.g. natural enemies.

Bolting: Premature shoot extension usually associated with the production of flowers and thus seeds. Usually due to environmental stimuli.

Bulb: A modified stem that served as an underground storage organ.

C_3 plant: Uses the Calvin cycle as the light independent phases of photosynthesis. Carbon dioxide combines with ribulose biphosphate to form to molecules of the three carbon phosphoglyceric acid.

C_4 plant: Light independent phase starts with the Hatch-Slack pathway which involves the use of the four carbon intermediate oxaloacetic acid. Most common in tropical plants. Overcomes the problems of photorespiration.

Cambium: A layer of actively dividing cells between the xylem and phloem.

Cardinal temperatures: Minimum, maximum and optimum temperatures at which important biochemical reactions within the plant occur.

Carpel: A leaflike structure containing the ovules.

Chromosome: Structures within the nucleus which contain the genes.

Clay soil: Soils containing a large proportion of clay particles (less than 0.002 mm in diameter).

Climacteric respiration: Some fruit show a burst of respiration rate before ripening.

Co-dominance: Neither form within a pair of genes is dominant to the other. This leads to a combining of characteristics to produce an intermediate form.

Cold requirement: Some species require a period of cold before they are able to flower and/or germinate.

Common Agricultural Policy (CAP): Legislation introduced by the EC in 1957. CAP aims to increase agricultural output, ensure a stable market and guarantee a fair price for produce and a fair standard of living for farm workers.

Corm: A short, enlarged base of a stem in which food is stored.

Crop rotation: System by which a farmer grows different crops in each of his fields each year in order to reduce the problems of disease.

Cross pollination: Pollination of flowers by pollen from another flower or another plant of the same species.

Cuttings: Parts of a plant taken during artificial vegetative propagation.

Cytokinins: A group of plant growth regulators.

Day neutral plants: The length of light or dark received by these plants does not stimulate a physiological response.

Detritus: Organic debris from decomposing plants and animals.

Dichogamy: Male and female parts of the flower mature at different times so self-pollination cannot occur.

Dicotyledon: Subclass of plants having two cotyledons at the first node of the stem.

Diploid: Both members of each pair of chromosomes are found within the cell. The normal condition for all body cells.

Dominance: Used in reference to the form of the gene pair that is expressed if present in an individual.

Dormant: Stage of development in a plant when metabolic processes are at a minimum, despite environmental conditions being conducive to growth.

Dry mass: The mass of material which has been dried at 110 °C until there is no further change in mass.

Dry matter: Mass of an object after drying to remove water.

DNA restriction endonuclease: Enzyme which slice DNA molecules into fragments.

DNA ligase: Enzyme which rejoins the ends of DNA fragments.

ECU: European Currency Unit.

EC: European Community also known as the European Economic Community (EEC).

Emasculation: The removal of the male parts of a plant. Used to prevent self pollination.

Embryo: Formed as a result of fertilisation of the ova by the pollen nucleus.

Embryo sac: The sac that contains the female gamete and seven other haploid nuclei.

Endosperm: Food storage material in a seed.

Epistasis: Used in reference to inheritance to describe the situation when one gene controls the expression of another.

ESA: Environmentally Sensitive Area.

Ethylene: A plant hormone involved in ripening.

Eutrophication: The enrichment of water by minerals (usually due to leaching or pollution) leading to the over production of plant matter.

Evapotranspiration: Combination of water loss by the processes of evaporation and transpiration.

Fertilisation: The union of male and female gametes to produce a zygote.

Fertiliser: General name given to organic or inorganic material added to the land to provide minerals for plant growth.

Field capacity: The amount of water that is held by a soil after it has been allowed to drain.

Floriculture: The commercial production of flowers.

Florigen: Name given to an as yet undiscovered, hormone or group of hormones which affect flowering.

Food chain: Chain of organisms in a natural community linked by their feeding relationship.

Frequency distribution: Diagram showing the frequency of a characteristic within a population.

Fruit: Produced from the ovary after fertilisation.

Fungicide: Chemical to kill fungi.

Gamete: Reproductive cell such as an ova or pollen grain nuclei. Contain haploid numbers of chromosomes.

Gene: Unit of inheritance made up of DNA and forming part of a chromosome.

Genotype: The genetic makeup of the nucleus.

Gene Transfer: The introduction of genes from the cells of one organism into the cells of another.

Germination: The sequence of events which occur in a seed leading to the outgrowth of the embryo, usually to the point of radicle protrusion.

Gibberellins: Plant hormones which control many aspects of plant development.

Grafting: A method of vegetative propagation which involves the transfer of parts of one plant onto parts of another and the fusion of their tissues.

Greenhouse: Protective environment for growing plants.

Green Revolution: The introduction of mechanisation, improved crop varieties and agrochemicals in an attempt to increase yields.

Growth regulator: Plant hormone, either natural or synthetic.

Haploid: A set of chromosomes containing only one of each pair of genes.

Herbicide: Chemical for killing weeds.

Heritability: The genetic component of the characteristics expressed by an individual.

Heterozygous: Containing both forms of the gene.

High yielding varieties: Varieties of rice and wheat which have been developed to produce higher yields in certain environmental conditions.

Homozygous: Containing two genes of the same form.

Hybrid: An organism produced as a result of the crossing of two different strains.

Humus: Organic fraction of the soil. Produced from the decayed bodies of animals and plants.

Inbred line: A highly homozygous organism produced as a result of continual self-pollination.

Inbreeding: Self-pollination.

Inbreeding depression: Loss of vigour due to continual inbreeding.

Incompatibility: A genetic condition where the male gametes are unable to pollinate certain stigmas.

Inflorescence: Group of flowers, usually associated with wind-pollinated flowers.

Intercropping: The growing of more than one crop in a field at a time.

Integrated Pest Management: The control of a pest population so that numbers are small enough not to cause uneconomic levels of damage.

Intervention: Scheme introduced by the EC in an attempt to reduce grain surpluses and thus control prices.

Insecticide: Chemical to kill insects.

Irrigation: Watering of crops, usually mechanical.

June Drop: Self thinning – a condition common in apples. Fruit which cannot compete successfully for carbohydrate during their growth will be aborted from the tree.

Juvenile phase: Phase of growth preceding maturity or the ability to become sexually reproductive.

Leaching: The washing out of soluble material. Usually used in reference to the removal of nitrates from the soil and their entry into the water system.

Leaf area index: Measurement of the ratio of leaf area to the ground occupied by a crop.

Leaf area duration: The time over which a crop has a particular leaf area.

Leaf-protein concentrate: Concentrate food supplement, rich in protein, minerals and vitamins, which is produced by the mechanical extraction of leaves.

Legume: A member of the pea family. Seeds are rich in protein as the roots of the plants support nitrogen fixing bacteria.

LFA: Less favourable areas.

Loam: A soil containing clay, silt and sand particles in proportions that produce a good water holding capacity in the soil.

Lodging: The falling down of a standing crop – usually associated with cereals.

Long day plants: Plants which flower only after they have been subjected to day light for more than a critical period of time in any 24 hours.

Mature phase: Stage of life cycle. Plant is able to flower and reproduce.

Meat analogue: Products which resemble meat but which are produced from vegetable proteins.

Mechanisation: The use of machines.

Meiosis: Cell division associated with gamete formation.

Meristem: Growing region or dividing cells.

Micropropagation: The production of new plants from small pieces of tissue or cells.

Minimum cultivation: Seed bed is prepared using the minimum levels of mechanisation. There is often an increased requirement for pesticides to destroy the weed that normal cultivation practices would remove.

Mitosis: Cell division associated with growth.

Monoculture: The continual production of the same crop in the same field.

Monocotyledon: Subclass of plants that have only one cotyledon at the first node of the stem.

Monoecious: Plants which have separate male and female flowers on the same plant e.g. maize.

Multiple genes: The control of one characteristic involving more than one pair of genes.

Net assimilation rate: Rate at which matter within a plant is assimilated, taking into consideration that used up during respiration etc.

Net primary production: Growth.

Outbreeding: Cross-pollination.

Parthenocarpy: Fruit production without sexual reproduction.

Passenger DNA: DNA fragment containing desired genes which is extracted from a cell during gene transfer.

Pathogen: Disease causing organism.

Pectin: A component of the plant cell wall.

Pectinase enzymes: Enzymes which hydrolyse the pectin in cell walls. Used in the production of some fruit juices.

Permanent wilting percentage: The minimum amount of water which is required within a soil in order to prevent permanent wilting.

Pesticide: Chemical to kill pests.

Phenotype: Characteristics which are shown by an individual.

Photoperiodic response: Responses to a plant of certain periods of light/dark within a 24 hour cycle.

Photorespiration: Type of respiration stimulated by light and common in certain species.

Photosynthesis: The production of food from inorganic material using light energy.

Phytochrome: Plant pigment, existing in interchangeable forms P_R and P_{FR}. Responsible for causing both flowering and germination in some species. Activated by certain wavelengths of light at low energy levels.

Pollination: The transfer of pollen to the stigma.

Polygene: A collection of genes which have an additive effect in the inheritance of certain characteristics.

Polyploidy: The multiplication of diploid numbers of chromosomes within a nucleus.

Precipitation: Rainfall.

Productivity: Yield. The rate at which energy is stored as matter.

Profile (soil): A section taken vertically through the soil which shows the layer or horizons from which it is composed.

Protandry: The maturation of male parts of a flower before the female parts so self pollination is avoided.

Protogyny: The maturation of the female parts of the flower before the male parts so that self pollination is avoided.

Pure line selective: A method of plant breeding involving the selection of desired traits and the inbreeding of individuals showing these traits.

Quota: Tonnage or hectarage allowance for a particular crop.

Relative growth rate: The rate at which a plant has increased in size per unit time, per unit of dry matter (i.e. relative increase in growth).

Recombinant DNA: DNA fragment formed when passenger and vector DNA combine.

Rootstock: The root onto which a scion is grafted.

Runners: A stem which grows horizontally above the ground as a method of vegetative spread and reproduction.

Sandy soil: A soil which contains a large proportion of sand particles (2–0.02 mm in diameter).

Scion: The shoot which is placed onto a root stock during grafting.

Self-pollination: The transfer of pollen from the anther of one flower to its own stigma or a stigma of another flower on the same plant.

Set aside: A scheme introduced by the EC in 1993 which aims to control the amount of combinable crops grown within Europe by paying subsidies to farmers to leave a percentage of their land fallow.

Share cropping: A worker provides the land owner with labour in return for a percentage of the product harvested.

Short day plants: Plants which will only flower if they receive less than the critical number of hours of light in a day or days.

Single-cell protein: Protein extracted for single cell organisms e.g. fungi, bacteria.

Somatic: Of the body (not a gamete cell).

Stratification: Cold treatment to stimulate seed germination.

Stolon: Overground stem used in reproduction.

Temph: A food product made from fermented soyabeans. Common in Japan.

Testa: Seed coat.

Tiller: Stem of grass plants.

Transpirational ratio: The amount of dry matter that is produced by a plant per kilogram of water absorbed.

Tropics: Area of the world which lies between the tropics of Cancer and Capricorn.

Tuber: Tip of an underground stem or root swollen with stored food.

Vegetative growth: Growth of shoots and roots. Growth concerned with non sexually reproductive structures.

Vegetative reproduction: Asexual reproduction.

Vernalisation: Cold treatment, stimulates flowering.

Weathering: The break down of rock to form soil by chemical and physical means.

Yield: The harvestable portion of the crop expressed in tonnes per hectare.

Zadoks Growth Key: A method of recognising stages of growth in some cereals.

INDEX

ACKNOWLEDGEMENTS

We are grateful to the following companies, institutions and individuals who have given permission to reproduce photographs in this book.

J. Allan Cash (1 bottom right, 2 all, 8 bottom right, 10 top, 18 top left, 32 top right, 42, 49, 68, 74 top, 81, 94, 104); Biophoto Associates (68 top left); Paul Brierly (131); Colin Taylor Productions (91 left, 92 left); Holt Studios (23 top right, bottom right, 33 top right, bottom, 34 both, 42 top, 47 left, 58 left, 61 left, 63 right, 64 all, 68, 69 top left, 73 top, 75 both, 79, 88 both, 89, 103 top, 120 left, 124 top, 127); ICI Agrochemicals (73 bottom); Leaf for Life Campaign (129 all four); Microscopix/ Andrew Syred (4 bottom left both); Roddy Paine (1 bottom left, top right, 9 top left, 47 right, 50 left, 90 both); Sutcliffe Electronics (69 bottom); Zeneca (63 left).

The artwork was drawn by Chartwell Illustrators.

We would also like to thank the following who gave permission to reproduce copyright material: Academic Press; Associated Examinations Board, English Nature; NIAB, Cambridge; Northern Examinations and Assessment Board; Oliver and Boyd Ltd. Plant Breeding International, Cambridge; University of Cambridge's Local Examinations Syndicate; University of Oxford Delegacy of Local Examinations; W.H. Freeman and Company Ltd.